VIII International Scientific Colloquium on Modelling for Materials Processing 2017

AA005653

IOP Conference Series: Materials Science and Engineering
Volume 355

Riga, Latvia
21 - 22 September 2017

Editors:

Andris Jakovics
Mihails Scepanskis

ISBN: 978-1-5108-6285-2
ISSN: 1757-8981

Printed from e-media with permission by:

Curran Associates, Inc.
57 Morehouse Lane
Red Hook, NY 12571

Some format issues inherent in the e-media version may also appear in this print version.

Copyright© (2017) by the Institute of Physics
All rights reserved. The material featured in this book is subject to IOP copyright protection, unless otherwise indicated.

Printed by Curran Associates, Inc. (2018)

For permission requests, please contact the Institute of Physics at the address below.

Institute of Physics
Dirac House, Temple Back
Bristol BS1 6BE UK

Phone: 44 1 17 929 7481
Fax: 44 1 17 920 0979

techtracking@iop.org

Additional copies of this publication are available from:

Curran Associates, Inc.
57 Morehouse Lane
Red Hook, NY 12571 USA
Phone: 845-758-0400
Fax: 845-758-2633
Email: curran@proceedings.com
Web: www.proceedings.com

Table of contents

Volume 355

VIII International Scientific Colloquium on Modelling for Materials Processing

21–22 September 2017, Riga, Latvia

Accepted papers received: 19 April 2018
Published online: 8 May 2018

Preface

Papers

Crystal growth

Metallurgical applications

EM Field & Heat treatment

Advanced applications

IOP Conf. Series: Materials Science and Engineering **355** (2018) 011001 doi:10.1088/1757-899X/355/1/011001

Dear distinguished Reader!

This is the proceeding of the VIII International Colloquium on Modelling for Material Processing, which took place on 21st and 22nd of September 2017 in Riga, Latvia. All papers within these proceedings are pre-reviewed and published as the issue of IOP Conference Series: Materials Science and Engineering. On behalf of the organizers of the colloquium, the University of Latvia and Leibniz University of Hanover, the editors of this issue kindly ask you to pay your attention on the scientific papers published in this issue.

The colloquium is the 8th among the series, which started in 1999 and were organized one by one in Riga, Latvia and Hanover, Germany. The success of this series of the colloquiums coincided with dramatic progress in numerical simulation. For example, if fluid dynamics, in 90s, scientists were trying to describe liquid flows in industrial metallurgical equipment with rough two-parametric models, while industry practiced purely experimental approach. With enhancement of computational resources, we saw transition in 2000s. Industry already adopted two-parametric simulation for averaged flows, while scientists were building reliable LES models. Nowadays, we see acceleration of knowledge transition between academic world and industry, LES came into practice in industrial R&D in 2010s. It was favoured by drastic jump in computer power and liberalization of simulation software market by open source initiatives.

This is just an example of increasing role of simulation in industry, physical prototyping has been replacing by virtual testing using simulation software to avoid design errors and reduce R&D costs and time to market. Taking into account the history successful knowledge transfer between academic world and industry in last decades, we believe the colloquiums like this play crucial role in industrial progress towards energy efficient and innovative manufacturing of parts and materials.

Traditionally, the series of the colloquiums reflected increasing role of electromagnetic technologies in smart manufacturing, including induction melting and stirring of electrically conductive materials. The majority of the papers reflects scientific progress in crystal growth and metallurgical industries. In many cases, the processes in multiphase materials with movement of phase interfaces and melting/ crystallisation are successfully researched. Nevertheless, the issue also includes some papers, which describes cutting-edge simulation applications in non-traditional way.

The editors hope that a reader will enjoy the scientific content of the issue, will find it useful for professional development and will explicitly cite these papers in the further works.

Editors

Dr. Phys. Andris Jakovics (University of Latvia)

Dr. Phys. Mihails Scepanskis (CENOS Ltd.)

Content from this work may be used under the terms of the Creative Commons Attribution 3.0 licence. Any further distribution of this work must maintain attribution to the author(s) and the title of the work, journal citation and DOI.
Published under licence by IOP Publishing Ltd

IOP Conf. Series: Materials Science and Engineering **355** (2018) 011002 doi:10.1088/1757-899X/355/1/011002

Peer review statement

All papers published in this volume of *IOP Conference Series: Materials Science and Engineering* have been peer reviewed through processes administered by the proceedings Editors. Reviews were conducted by expert referees to the professional and scientific standards expected of a proceedings journal published by IOP Publishing.

Content from this work may be used under the terms of the Creative Commons Attribution 3.0 licence. Any further distribution of this work must maintain attribution to the author(s) and the title of the work, journal citation and DOI.
Published under licence by IOP Publishing Ltd

IOP Conf. Series: Materials Science and Engineering **355** (2018) 012001 doi:10.1088/1757-899X/355/1/012001

The synergy of modeling and novel experiments for melt crystal growth research

Jeffrey J. Derby

Department of Chemical Engineering and Materials Science, University of Minnesota, 421 Washington Avenue, SE, Minneapolis, MN, 55455-0132, USA

E-mail: derby@umn.edu

Abstract. Computational modeling and novel experiments, when performed together, can enable the identification of new, fundamental mechanisms important for the growth of bulk crystals from the melt. In this paper, we present a compelling example of this synergy via the discovery of previously unascertained physical mechanisms that govern the engulfment of silicon carbide particles during the growth of crystalline silicon.

1. Introduction

The engulfment of foreign particles during the growth of multicrystaline silicon is a problem during the production of high-quality, low-cost photovoltaic devices. In particular, carbon contamination of the melt, from the original silicon feed and via gas phase reaction and transport from hot furnace internals, leads to the formation of silicon carbide (SiC) particles which may be engulfed during crystal growth and adversely affect the subsequent processing of solar cells [1].

These issues have motivated experimental analyses of particle incorporation during the directional solidification of silicon ingots [2, 3]. A common conclusion of these studies was that classical theories for engulfment were unable to describe the behavior of SiC engulfment. Since melt convection significantly affects engulfment, microgravity experiments have been performed in conjunction with terrestrial studies [4, 5]. These experiments have provided the best data yet available for SiC-Si engulfment; however, accompanying theoretical analyses were still unable to provide a real quantitative correlation with the experimental results [4, 6].

In the following discussion, we show rigorous numerical computations that provide a quantitative representation of the experimental results for this system via a predicted scaling of critical engulfment velocity with particle radius of $v_c \sim R^{-5/3}$. Prior, classical engulfment theories predict either $v_c \sim R^{-1}$ or $v_c \sim R^{-4/3}$. Informed by our computational results, we further describe a simple analysis that explains these scalings.

2. Experiments

Figure 1 illustrates the ParSiWal (Partikeleinfang bei der Siliziumkristallisation im Weltall) project, which was designed to study the engulfment of SiC particles during the solidification of silicon [4,5]. The experiments were conducted via the seeding of a silicon rod with pre-synthesized SiC particles (Figure 1a), which was then placed in a mirror furnace (Figure 1b) and flown in a low-orbit sounding rocket (Figure 1c) to achieve solidification and engulfment under microgravity conditions. The same growth system was also employed on earth for additional studies.

Content from this work may be used under the terms of the Creative Commons Attribution 3.0 licence. Any further distribution of this work must maintain attribution to the author(s) and the title of the work, journal citation and DOI.
Published under licence by IOP Publishing Ltd

Figure 1. Microgravity and terrestrial experiments measure critical conditions for SiC particle engulfment during silicon growth [4, 5]. Right: Silicon sample, seeded with SiC particles, in crucible. Middle: ELLI mirror furnace in which growth is conducted. Right: TEXUS 51 rocket flight to achieve microgravity experimental conditions.

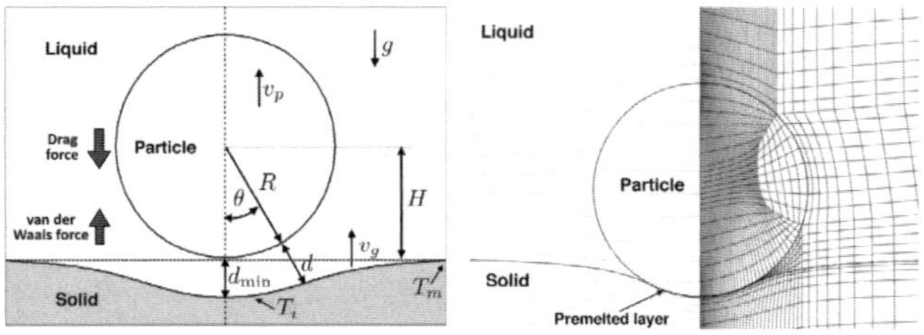

Figure 2. (a) A schematic depiction of particle engulfment during solidification. (b) A computation of a SiC particle article being engulfed by a silicon solidification front by our numerical model. Geometry and temperature isotherms are shown on the left, with the corresponding finite element mesh shown on the right.

These experiments have provided the best data yet available for SiC-Si engulfment; however, accompanying theoretical analyses were still unable to provide a quantitative description of experimental results [4, 6].

3. Model and experimental results

We have developed a rigorous numerical model for engulfment that is free from most of the simplifying assumptions that have been employed in previous approaches [7] and have reported steady-state and dynamic studies of engulfment during crystal growth in [8–10].

Our model describes a solid particle interacting with a solidification front, as depicted in Figure 2(a), with a typical example of a computation of a SiC particle by silicon shown in Figure 2(b). The crystal solidifies upward with a growth rate v_g, and, as melt-solid interface approaches the particle, repulsive van der Waals forces push it away from the interface, imparting a particle velocity v_p. Due to the upward particle motion, liquid must flow into the gap to fill the volume previously occupied by the particle, which results in hydrodynamic drag. If the drag forces—which hinder the particle from moving together with the advancing solid-liquid interface—dominate, the particle will be engulfed into crystal. Thus, a force balance on the particle is the pivotal step in modeling its engulfment. We will return to this force balance to explain the predicted scalings, by different models, of critical engulfment velocity with respect to particle size.

The experimental data and the results of a series of computations of engulfment are presented in Figure 3, which shows growth velocity versus particle diameter in a log–log plot. The terrestrial experimental conditions resulting in engulfed particles are represented by the solid, red squares, whereas pushed particles (not engulfed) are shown as open, green circles. The

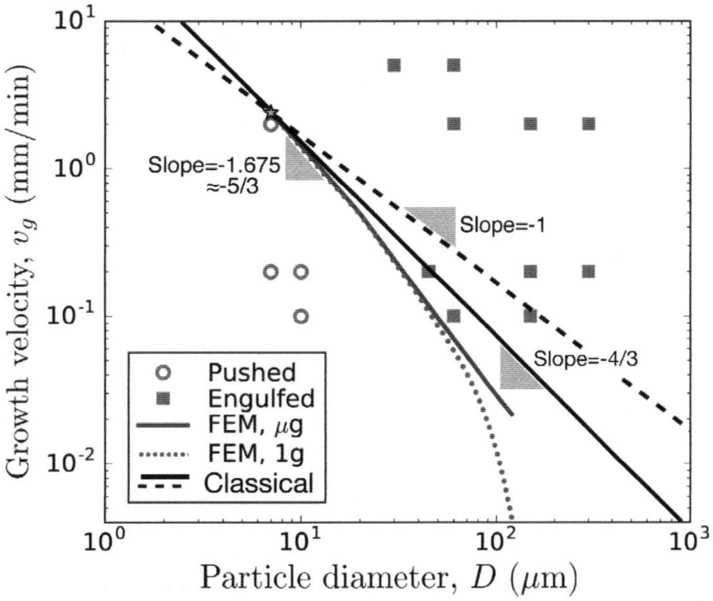

Figure 3. Experimental data and engulfment models for the SiC-Si system [10]. The symbols denote data from terrestrial experiments; the star denotes a high-confidence engulfment observation from a microgravity experiment. Curves show predictions of the critical velocity and mark the boundary between particle pushing (to the left and below) and engulfment (to the right and above). The black curves represent prior, classical models for engulfment. The results from our finite-element model (FEM) are shown by the blue, solid curve and the pink, dotted curve. The position and slope of the curves are discussed in the text.

single star indicates a high-confidence observation of an engulfed particle from a microgravity experiment.

The curves in Figure 3 indicate various model predictions for the critical velocity of engulfment, v_c, as functions of particle size. All have been calibrated to intersect the microgravity datum (star) to account for the unknown value of the Hamaker constant for this system, which sets the magnitude of the van der Waals repulsive force. These critical velocity curves should therefore separate the data points representing engulfed and pushed particles.

The two black lines show predictions for all prior, classical models of engulfment. The dashed, black line exhibits a slope of -1 and represents models that predict a scaling of critical velocity, as $v_c \sim R^{-1}$, such as those described in [2,11,12]. The solid, black line represents models that have more rigorously included Gibbs-Thomson curvature effects and pre-melting on the solidification interface [13,14] and predict a straight line with a slope of -4/3, thereby giving rise to a scaling of $v_c \sim R^{-4/3}$. Significantly, these classical models are not able to quantitatively represent the experimental data for this system; namely, the curves do not divide the data points representing engulfed and pushed states.

In contrast, our finite-element numerical model, shown in Figure 3 by the solid, blue curve (for microgravity) and the dotted, pink curve (including terrestrial gravitational effects), is able to quantitatively correlate these data. Namely, all of data points of the engulfed states lie above these curves, while pushed states (particles not engulfed) lie below them. Such agreement between theory and experiment has heretofore been unattainable for this silicon carbide–silicon system.

The success of our numerical model in describing these data is due to its steeper slope, measured as approximately −1.675, from which we deduce a scaling of $v_c \sim R^{-5/3}$. Significantly, this scaling has not, to our knowledge, been observed nor predicted in prior studies of engulfment. While we have addressed the origin of this scaling in some detail in a prior work [10], we present some simpler arguments below.

4. Analytical scaling arguments

Under critical engulfment conditions, the pushing force on the particle arising from van der Waals effects is just balanced by drag forces. We approximate these as:

$$\frac{2AR^3}{3d_{min}^2(2R+d_{min})^2} = \frac{6\pi\mu R^2 v_c}{d_{min}}, \tag{1}$$

where the left-hand side represents Hamaker's solution [15] for the van der Waals force and the right-hand side is the drag computed by lubrication theory [16], both derived for a sphere approaching a planar interface. Here, we represent the particle radius by R, the minimum gap thickness by d_{min}, and the critical solidification and particle velocity as $v_g = v_p = v_c$ (see Figure 2); A is the Hamaker constant, and μ is the melt viscosity.

If we consider that the particle radius is much greater than the minimum gap thickness, $R \gg d_{min}$, then the above expression can be simplified and rearranged to show the dominant scaling behavior:

$$\frac{R}{d_{min}^2} \sim \frac{R^2 v_c}{d_{min}}, \tag{2}$$

$$\Longrightarrow v_c \sim (Rd_{min})^{-1}. \tag{3}$$

The simplest classical theories for engulfment [2, 11, 12] assume that the gap between the particle and solidification interface is determined solely by molecular interactions, in which case d_{min} is independent of the particle radius, giving, via Eq. (3), a scaling of $v_c \sim R^{-1}$. More rigorous classical models that included Gibbs-Thomson curvature and pre-melting on the solidification interface [13, 14] predict that $d_{min} \sim R^{1/3}$. With this result, Eq. (3) predicts $v_c \sim R^{-4/3}$.

Our computations for the SiC-Si system [10] showed a significant, previously unappreciated interaction between particle-induced deflection of the solidification interface, via the effects of heat channeling through the particle when its thermal conductivity is greater than that of the melt, and a flattening of the most curved portion of the interface, via the Gibbs-Thomson effect. Thus, for a range of particle sizes between approximately 10–100 μm in this system, the solidification interface takes on a rather complicated shape. We argue that this complicated interface shape has different effects on the van der Waals repulsion and hydrodynamic drag, leading to a reinterpretation of the force balance of Eq. (2) as:

$$\frac{R}{d_{vdW}^2} \sim \frac{R^2 v_c}{d_D}, \tag{4}$$

$$\Longrightarrow v_c \sim \frac{d_D}{Rd_{vdW}^2}, \tag{5}$$

where we have introduced different gap thicknesses d_{vdW} and d_D for computation of the van der Waals and drag forces, respectively.

We argue that the short-range nature of the van der Waals force makes it most sensitive to the minimum approach distance, so that $d_{vdW} \approx d_{min}$ and, following prior scaling arguments [13,14], $d_{vdW} \sim R^{1/3}$. However, the net hydrodynamic drag is computed via an integration along the

IOP Conf. Series: Materials Science and Engineering **355** (2018) 012001 doi:10.1088/1757-899X/355/1/012001

gap between the particle surface and the complicated interface shape. The size of this gap varies, made larger in general by the heat flow through the particle but decreased in thickness at the centerline, where the Gibbs-Thomson effect flattens the deepest, most curved portion of the interface; see [10] for more details. We posit that the outcome is a drag force that does not significantly change with gap, so that d_D can be considered to be constant with respect to R. Use these scalings for d_{vdW} and d_D in Eq. (5) leads to the observed behavior of $v_c \sim R^{-5/3}$.

5. Conclusions

Motivated by the inability of classical models to describe experimental data on the engulfment of SiC during silicon crystal growth, we have developed a rigorous, numerical model to study this system. For the first time in over a decade of research on SiC inclusions in silicon, our model is able to provide a quantitative correlation of experimental results, and we are able to unequivocally identify the underlying physical interactions that give rise to the observed behavior. In particular, we identified a significant and previously unascertained interaction between particle-induced interface deflection, originating from the thermal conductivity of the SiC particle being larger than that of the surrounding silicon liquid, and curvature-induced changes in melting temperature arising from the Gibbs-Thomson effect. The interface shapes arising from these interactions produce different effects on the van der Waals and hydrodynamic drag forces and result in a scaling of $v_c \sim R^{-5/3}$ for particle sizes of relevance in this system. This synergy of modeling and experiments has thus provided new insight into the fundamentals of crystal growth.

Acknowledgments

Acknowledgments This research was supported in part by NASA NNX10AR70G; no official endorsement should be inferred.

References

[1] Du G, Chen N, and Rossetto P 2008 On-wafer investigation of SiC and Si_3N_4 inclusions in multicrystalline Si grown by directional solidification, *Solar Energy Materials and Solar Cells* **92** (9) 1059–1066.

[2] Søiland A K, Øvrelid E J, Engh T, Lohne O, Tuset J K, and Gjerstad O 2004 SiC and Si_3N_4 inclusions in multicrystalline silicon ingots, *Materials Science in Semiconductor Processing* **7** 39–43.

[3] Trempa M, Reimann C, Friedrich J, and Muller G 2010 The influence of growth rate on the formation and avoidance of C and N related precipitates during directional solidification of multi crystalline silicon, *Journal of Crystal Growth* **312** (9) 1517–1524.

[4] Friedrich J, Reimann C, Jauss T, Cröll A, and Sorgenfrei T 2016 Interaction of SiC particles with moving solid–liquid interface during directional solidification of silicon, *Journal of Crystal Growth* **447** 18–26.

[5] Sorgenfrei T, Jauss T, Cröll A, Reimann C, Friedrich J, and Volz M 2017 The critical growth rate for particle incorporation during the directional solidification of solar silicon under microgravity, *Int. J. Microgravity Sci. Appl.* **34** 340115-1–8.

[6] Aufgebauer H, Kundin J, Emmerich H, Azizi M, Reimann C, Friedrich J, Jauss T, Sorgenfrei T, and Cröll A 2016 Phase-field simulations of particle capture during the directional solidification of silicon, *Journal of Crystal Growth* **446** 12–26.

[7] Tao Y, Yeckel A, and Derby J J 2016 Steady-state and dynamic models for particle engulfment during solidification, *J. Comp. Phys.* **315** 238–263.

[8] Tao Y, Yeckel A, and Derby J J 2016 Analysis of particle engulfment during the growth of crystalline silicon, *Journal of Crystal Growth* **452** 1–5.

[9] Tao Y, Sorgenfrei T, Jauss T, Cröll A, Reimann C, Friedrich J, and Derby J J 2016 Particle engulfment dynamics under oscillating crystal growth conditions, *Journal of Crystal Growth* **468** 24–27.

[10] Derby J J, Tao Y, Reimann C, Friedrich J, Jauss T, Sorgenfrei T, and Croell A 2017 A quantitative model with new scaling for silicon carbide particle engulfment during silicon crystal growth, *Journal of Crystal Growth* **463** 100–109.

[11] Uhlmann D R, Chalmers B, and Jackson K A 1964 Interaction between particles and a solid-liquid interface, *J. Appl. Phys.* **35** 2986.

[12] Stefanescu D M, Juretzko F R, Dhindaw B K, Catalina A, Sen S, and Curreri P A 1998 Particle engulfment and pushing by solidifying interfaces: Part II. Microgravity experiments and theoretical analysis, *Metall. and Mat. Transactions A* **29A** 1697.

[13] Chernov A A and Temkin D E 1976 Capture of inclusions in crystal growth, in: Kaldis E and Scheel H J (Eds.), *Crystal Growth and Materials, Vol. 2 of Current Topics in Materials Science,* North-Holland Publishing Company, Amsterdam, pp. 3–77.

[14] Rempel A W and Worster M G 2001 Particle trapping at an advancing solidification front with interfacial-curvature effects, *J. Crystal Growth* **223** 420–432.

[15] Hamaker H C 1937 The London–van der Waals attraction between spherical particles, *Physica* **IV** (10) 1058–1072.

[16] Cox R G and Brenner H 1967 The slow motion of a sphere though a viscous fluid towards a plane surface—II Small gap widths, including inertial effects, *Chem. Eng. Sci.* **22** 1753–1777.

IOP Conf. Series: Materials Science and Engineering **355** (2018) 012002 doi:10.1088/1757-899X/355/1/012002

Modeling Czochralski growth of oxide crystals for piezoelectric and optical applications

C Stelian and T Duffar

Université Grenoble Alpes, CNRS, Grenoble INP, SIMAP, F-38000 Grenoble, France

Abstract. Numerical modeling is applied to investigate the impact of crystal and crucible rotation on the flow pattern and crystal-melt interface shape in Czochralski growth of oxide semi-transparent crystals used for piezoelectric and optical applications. Two cases are simulated in the present work: the growth of piezoelectric langatate (LGT) crystals of 3 cm in diameter in an inductive furnace, and the growth of sapphire crystals of 10 cm in diameter in a resistive configuration. The numerical results indicate that the interface shape depends essentially on the internal radiative heat exchanges in the semi-transparent crystals. Computations performed by applying crystal/crucible rotation show that the interface can be flattened during LGT growth, while flat-interface growth of large diameter sapphire crystals may not be possible.

1. Introduction

Czochralski (Cz) technique is used for the growth of a variety of oxide materials such as Al_2O_3 (sapphire), $Y_3Al_5O_{12}$ (YAG), $LiNbO_3$, $Bi_4Ge_3O_{12}$ and $La_3Ga_{5.5}Ta_{0.5}O_{14}$ (LGT). Among these oxides, LGT and his homologous compounds, are of interest as raw materials for piezoelectric applications. Sapphire has also many applications, being used for lasers, optical windows or as substrate for Light Emitting Diodes (LED) devices.

The Czochralski method has been well developed in time for the growth of oxide crystals [1]. However, the quality of crystals is still affected by defects as dislocations, inclusions, cracks and facets. Numerical simulation has become an important tool to investigate Cz growth of semi-transparent oxide crystals [2-8]. In their work [2], Xiao and Derby studied the effect of the internal radiation in Cz growth of yttrium aluminium garnet (YAG) and gadolinium gallium garnet (GGG). They found that the conical shape of the crystal-melt interface depends on the internal radiant transfer through the crystal. Other papers investigated the heat and mass transport phenomena at different stages of the Cz growth process [3, 4]. It is generally recognized that the deeply convex shape of the interface in these oxide systems, generate high thermal stress in the crystal, which may promote cracking, facets and dislocations [5, 6]. Crystal and crucible rotation can be applied to flatten the interface and reduce thermal stresses [2, 7, 8]. Numerical computations carried out in [7, 8] have shown that the interface can be flattened in Cz sapphire growth by applying crystal/crucible counter rotation. However, no experimental validation of these numerical results has been provided. It should be pointed that in practice, only crystal rotation is used in Czochralski growth of oxides.

The objective of the present work is to investigate the impact of crystal and crucible rotation on the flow pattern and interface shape in Cz growth of LGT and sapphire crystals.

Content from this work may be used under the terms of the Creative Commons Attribution 3.0 licence. Any further distribution of this work must maintain attribution to the author(s) and the title of the work, journal citation and DOI.
Published under licence by IOP Publishing Ltd

IOP Conf. Series: Materials Science and Engineering **355** (2018) 012002 doi:10.1088/1757-899X/355/1/012002

2. Model description

Numerical modeling is performed by using the finite element code COMSOL Multiphysics. Heat transfer and momentum equations are solved for a 2D-axisymmetric furnace configuration, where all the furnace components are included. The internal radiative heat transfer in semi-transparent LGT and sapphire crystals is simulated by using P1-approximation model. The laminar flow computations include buoyancy, Marangoni and forced convection due to crystal and crucible rotation. Simulations are performed at a given stage of the crystallization process. A model which uses the deformable mesh technique to compute the shape of the crystal-melt interface, has been implemented in COMSOL Multiphysics. Transient computations with deformed mesh are conducted by using initial conditions for temperatures and flow velocities carried out from steady-state simulations.

3. Numerical results

The modeling is performed for two different Czochralski configurations. LGT crystals of 3 cm in diameter are grown in an inductive heated furnace (see Fig. 1). The electromagnetic field produced by a RF coil induces eddy currents in an iridium crucible (6 cm in diameter) which is heated by Joule's effect. Figure 1 shows the temperature field (at the left side), streamlines and flow velocity (at the right side) computed at the middle stage of the growth process. The solidification isotherm is shown (in blue colour) on the figure. High temperatures, above the melting point are obtained only in the melt and a small region around the crucible. The radial temperature gradient at the free surface of the melt, which drives the Marangoni convection, is $G_{Tr} = 200\text{K/cm}$.

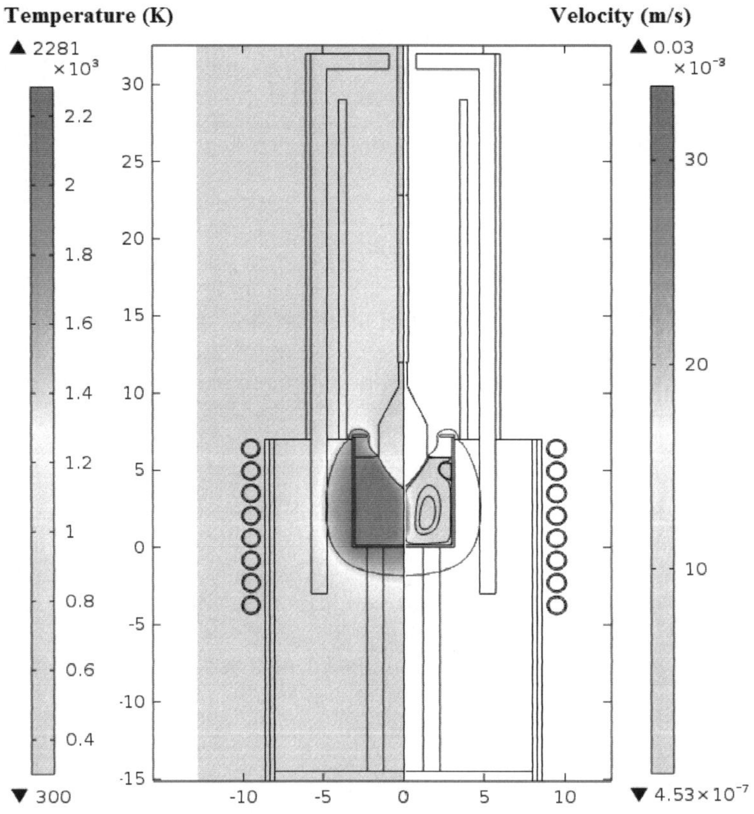

Figure 1. Temperature field (left side), streamlines and velocity field (right side) computed in an inductive heated Czochralski configuration used to grow LGT crystals of 3 cm in diameter.

Figure 2 shows a resistive heated Czochralski configuration used to grow sapphire crystals of 10 cm in diameter. The input power in the heater is much higher as compared to the previous case. Numerical computations show a more extended area of high temperatures in the furnace (see the left side of the Fig. 2). It is found that the resistive heating ensures small temperature gradients in the melt. The radial temperature gradient at the free surface of the melt is $G_{Tr} = 28K/cm$.

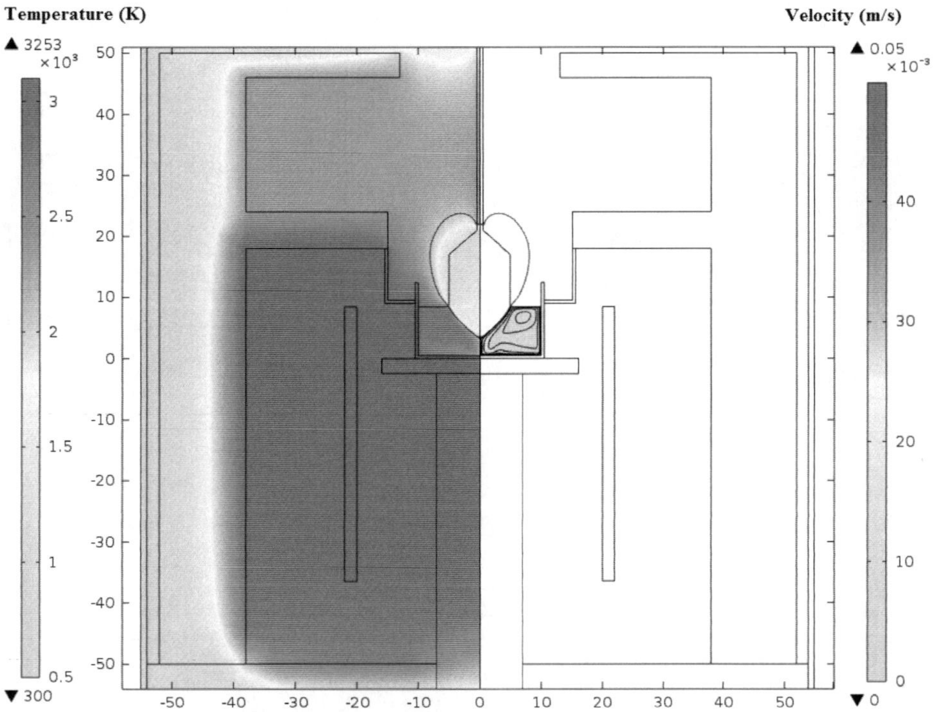

Figure 2. Temperature field (left side), streamlines and velocity field (right side) computed in a resistive heated Czochralski configuration used to grow sapphire crystals of 10 cm in diameter.

3.1. Internal radiative effect

The effect of the radiative heat exchanges in the semi-transparent LGT crystal is analysed in Fig. 3. Figure 3a shows the results carried out from a computation which takes into account the internal radiative heat exchanges in the crystal. The radiation is absorbed and emitted by the semi-transparent LGT ingot, leading to a significant increase of the effective thermal conductivity of the crystal. The huge difference between the thermal conductivities in the solid and liquid phases is responsible for the deeply convex shape of the interface. The computed interface deflection ($f \cong 2.2cm$) was validated by comparison to experimental measurements in our previous work [9]. The flow pattern is characterized by one main vortex which flows upwards along the crucible wall and downwards at the sample centreline. The interface becomes almost flat in the computation performed for an opaque crystal (Fig. 3b). The radial temperature difference in the crystal at the proximity of the triple solid-liquid-gas point is much higher in the case which accounts for the internal radiative effect: $\Delta T = 110K$ (semi-transparent crystal) and $\Delta T = 15K$ (opaque crystal).

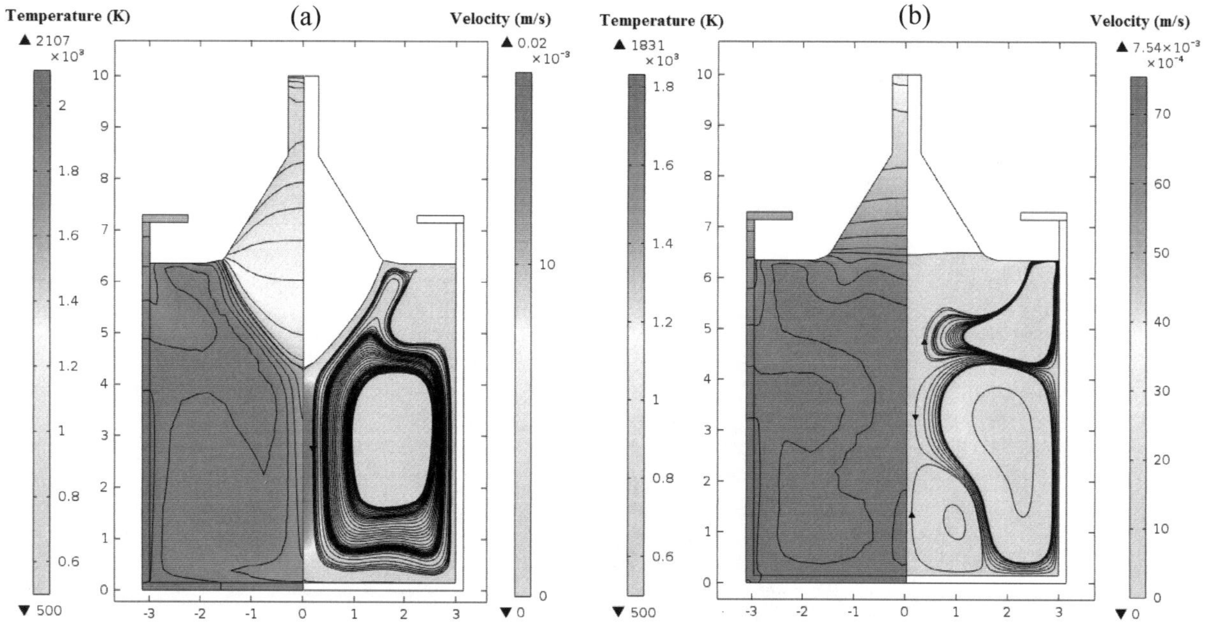

Figure 3. Temperature field and isotherms (left side), streamlines and velocity field (right side) computed for Cz growth of LGT: (a) semi-transparent crystal; (b) opaque crystal.

3.2. Effect of crystal and crucible rotation

The effect of crystal and crucible rotation in Cz growth of piezoelectric LGT crystals of 3 cm in diameter was analysed in our previous work [9]. Numerical computations have shown that the interface can be flattened by applying only crystal rotation at high rates ($\omega_{crystal} = 45 - 50\text{rpm}$), due to reversed flow at the sample centreline underneath the crystal. Applying crystal/crucible counter rotation in this configuration, leads to a distorted shape of the interface.

The effect of the forced convection in the case of sapphire crystals (10 cm in diameter) is analysed in Fig. 4. Computations are performed for a crystal length of L = 4 cm (conical part of the ingot). The case computed without rotation shows that the liquid downwards at the sample centre, below the crystal (Fig. 4a). The computed interface deflection ($f \cong 5.3\text{cm}$) is in agreement with the experimental result in [10]. Applying only crystal rotation at rates greater than a critical value ($\omega_{crystal} = 20\text{rpm}$), changes the flow direction at the sample centreline. The upward flow located below the crystal has a flattening effect on the growth interface. However, this effect becomes important starting with rotation rates of $\omega_{crystal} = 30\text{rpm}$ (Fig. 4b). In this case, the interface curvature decreases, but the convection becomes unstable, having a deleterious effect on the interface shape, which is slightly distorted. If the rotation rate is increased above 30rpm, the interface is more distorted by the flow. Figure 4c shows the case computed with crystal and crucible counter rotation: $\omega_{crystal} = 30\text{rpm}$, $\omega_{crucible} = -30\text{rpm}$. Applying crucible rotation intensifies the downward flow below the crystal, and opposes the beneficial effect of the crystal rotation, so the interface remains curved.

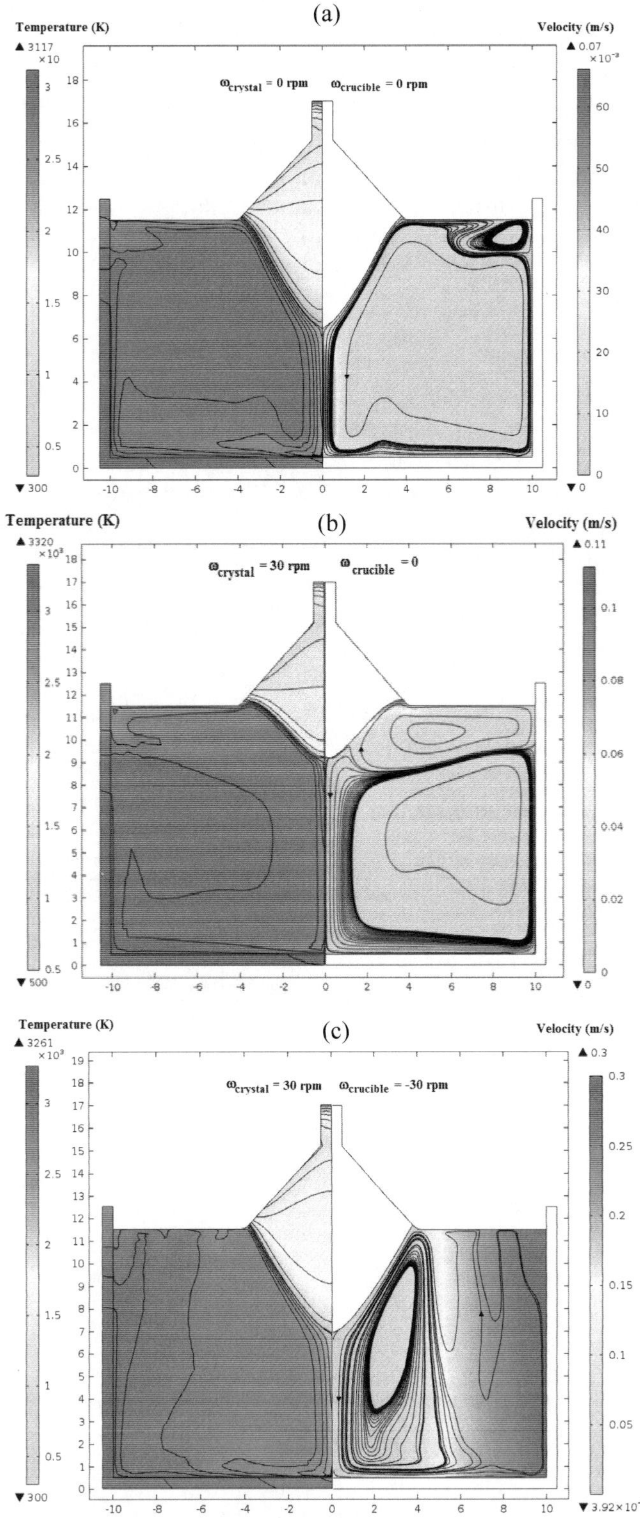

Figure 4. Temperature field and isotherms (left side), streamlines and velocity field (right side) computed with crystal/crucible rotation in the case of sapphire growth: (a) $\omega_{crystal} = 0$, $\omega_{crucible} = 0$; (b) $\omega_{crystal} = 30\text{rpm}$, $\omega_{crucible} = 0$; c) $\omega_{crystal} = 30\text{rpm}$, $\omega_{crucible} = -30\text{rpm}$.

4. Conclusion

Previous numerical analysis of crystal/crucible rotation effect in Czochralski growth of small diameter (3 cm) piezoelectric LGT crystals have shown that the interface can be flattened by applying only crystal rotation at high rates $\omega_{crystal} = 45 - 50\text{rpm}$ [9]. The present computations show that the growth interface cannot be flattened in the case of large size sapphire crystals (10 cm in diameter). Applying only crystal rotation at high rates results in a distorted shape of the interface, while crystal/crucible counter rotation has no flattening effect, due to the opposite directions of the induced flows. These results are in agreement with experimental Cz growth of sapphire crystals, which generally uses only crystal rotation at low rates [6].

5. References

[1] Brandle C D 2004 *J. Cryst. Growth* **264** 593
[2] Xiao Q and Derby J J 1994 *J. Cryst. Growth* **139** 147
[3] Tavakoli M H 2008 *J. Cryst. Growth* **310** 3107
[4] Lu C W, Chen J C, Chen C H, Chen, C H, Hsu W C and Liu C M 2010 *J. Cryst. Growth* **312** 1074
[5] Tsukada T, Kakinoki K, Hozawa M, Imaishi N, Shimamura K and Fukuda T 1997 *J. Cryst. Growth* **180** 543
[6] Fang H S, Pan Y Y, Zheng L L, Zhang Q J, Wang S and Jin Z L 2013 *J. Cryst. Growth* **363** 25
[7] Hur M J, Han X F, Song D S, Kim T H, Lee N J, Jeong Y J and Yi 2014 *J. Cryst. Growth* **385** 22
[8] Nguyen T P, Hsieh Y T, Chen J C, Hu C and Nguyen HB 2017 *J. Cryst. Growth* **468** 514
[9] Stelian C, Nehari A, Lasloudji I, Lebbou K, Dumortier M, Cabane H and Duffar T 2017 *J. Cryst. Growth* **475** 368
[10] Bruni F J 2015 *Cryst. Res. Technol.* **50** 133

IOP Conf. Series: Materials Science and Engineering **355** (2018) 012003 doi:10.1088/1757-899X/355/1/012003

Modelling of thermal field and point defect dynamics during silicon single crystal growth using CZ technique

A Sabanskis and J Virbulis

Department of Physics, University of Latvia, Zellu str. 23, LV-1002, Riga, Latvia

E-mail: andrejs.sabanskis@lu.lv, janis.virbulis@lu.lv

Abstract. Mathematical modelling is employed to numerically analyse the dynamics of the Czochralski (CZ) silicon single crystal growth. The model is axisymmetric, its thermal part describes heat transfer by conduction and thermal radiation, and allows to predict the time-dependent shape of the crystal-melt interface. Besides the thermal field, the point defect dynamics is modelled using the finite element method. The considered process consists of cone growth and cylindrical phases, including a short period of a reduced crystal pull rate, and a power jump to avoid large diameter changes. The influence of the thermal stresses on the point defects is also investigated.

1. Introduction

The Czochralski method is widely used to produce silicon single crystals for the semiconductor industry. The crystal quality depends on the microdefect distribution, which is determined by the self-interstitials and vacancies, as well as the dopants. During the process, two PID controllers are used to adjust the crystal pull rate and the heater power [1], therefore, the whole CZ process is non-stationary (transient), although a steady-state approach is often applied, e.g., for the cylindrical growth stage. The dynamical nature of the CZ process is especially important for the point defect calculations, due to the dependence on the thermal history of the crystal.

In the present paper, the existing model [2] is extended with the possibility to calculate the point defect distribution and thermal stresses in the crystal. Transient calculations of cone and cylindrical growth are carried out, including a pull rate drop in cylindrical stage. The novelty of the present work is that the power jump and power control is explicitly considered in the simulations to avoid large diameter changes.

2. Mathematical model

2.1. Temperature and phase boundaries

An axisymmetric geometry is considered, see Fig. 1. The non-stationary temperature field and the changing shape of the phase boundaries are calculated with the program `CZ-Trans` [2]. Heat transfer by conduction and thermal radiation is considered, while the influence of the convective heat transfer is modelled indirectly via increased thermal conductivity of the melt. Instead of the absolute heater power, the power change ΔP is used in the model. The heater is described as a surface with a homogeneous temperature distribution which changes in time; the model takes into account ΔP, net thermal radiation, heat losses and heat capacity [2].

Content from this work may be used under the terms of the Creative Commons Attribution 3.0 licence. Any further distribution of this work must maintain attribution to the author(s) and the title of the work, journal citation and DOI.
Published under licence by IOP Publishing Ltd

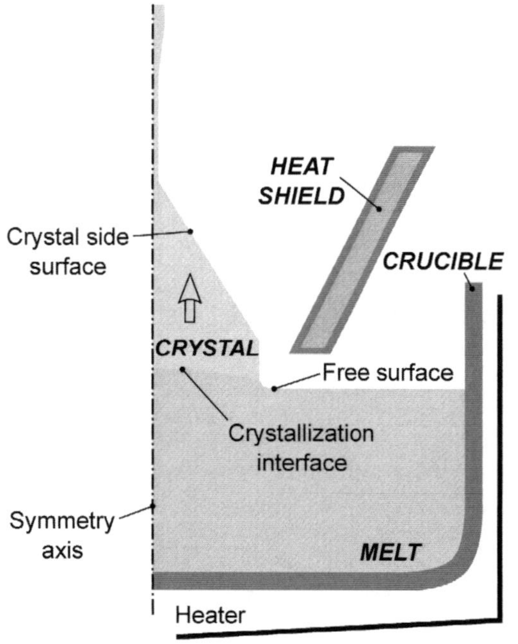

Figure 1. Scheme of the considered CZ system (in scale). Crystal diameter 200 mm.

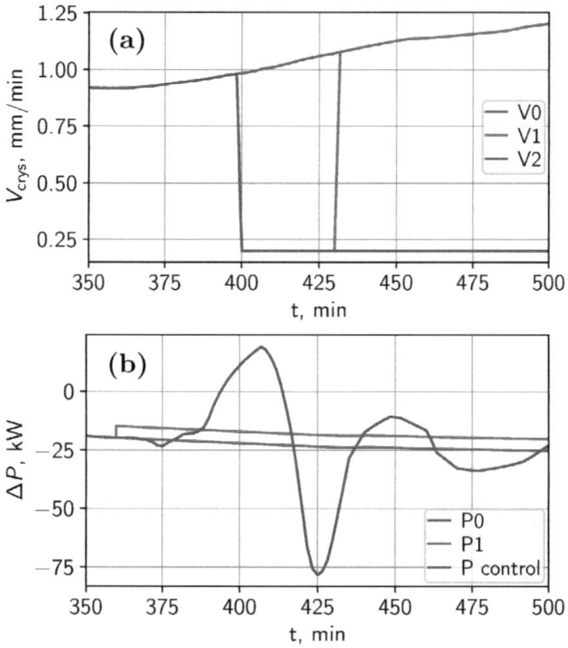

Figure 2. Time-dependence of (a) crystal pull rate and (b) heater power change.

2.2. Point defects

Two types of the point defects in the crystal are considered: self-interstitials, denoted by the index "I", and vacancies, denoted by the index "V". The corresponding concentrations C_I and C_V [cm^{-3}] are coupled by the recombination term – the last term in the advection-diffusion-reaction equation (1). The subscript "I,V" is a shorthand, meaning that (1) represents two equations – the first one for I and the second one – for V. The thermodiffusion term is omitted since the corresponding physical parameters are not precisely known and its estimated influence is much weaker than the effect of the equilibrium concentrations [3].

$$\frac{\partial C_{I,V}}{\partial t} + \vec{u}\nabla C_{I,V} = \nabla\left(D_{I,V}\nabla C_{I,V}\right) + k_{\text{rec}}\left(C_I^{\text{eq}}C_V^{\text{eq}} - C_I C_V\right). \tag{1}$$

In the equation (1), t is the time, $\vec{u}$ is the velocity (crystal pull rate V_{crys}), C^{eq} is the equilibrium concentration, D is the diffusion coefficient [m^2/s], and k_{rec} is the reaction constant for recombination [cm^3/s]. On the symmetry axis (see also Fig. 1), the zero flux boundary condition (BC) is set: $\partial C_{I,V}/\partial n = 0$. At the crystallization interface and the crystal side surface, the equilibrium concentrations are applied: $C_{I,V} = C_{I,V}^{\text{eq}}$.

Once the concentrations C_I and C_V are calculated, the parameter $\Delta = C_I - C_V$ can be used to analyse the excess of the self-interstitials/vacancies in different regions of the crystal. Zero isoline of Δ is the position of the I-V boundary.

The physical parameters for point defects strongly depend on the absolute temperature, T. The general expressions for the diffusion coefficient and the equilibrium concentration are given by the equation (2), where $k = 8.617\cdot10^{-5}\,\text{eV/K}$ is the Boltzmann constant, and the parameters D^0, E^m, C^0, H^f and S^f are taken from [4].

$$D(T) = D^0 \exp\left[-\frac{E^m}{kT}\right], \quad C^{\text{eq}}(T) = C^0 \exp\left[-\frac{H^f - S^f T}{kT}\right]. \tag{2}$$

At the melting point of silicon, $T_0 = 1685\,\mathrm{K}$, the values are $C_I^{\mathrm{eq}} = 6.9 \cdot 10^{14}\,\mathrm{cm}^{-3}$, $C_V^{\mathrm{eq}} = 8.5 \cdot 10^{14}\,\mathrm{cm}^{-3}$, and $\Delta = -1.6 \cdot 10^{14}\,\mathrm{cm}^{-3}$, which means that at the crystallization interface more vacancies are formed than interstitials. Nevertheless, if the crystal pull rate is low enough, further away from the interface the crystal can be interstitial-rich since at high temperatures $D_I > D_V$, resulting in a stronger diffusive flux of self-interstitials [5].

The recombination coefficient is given by the equation (3), where the capture radius for recombination $a_r = 1\,\mathrm{nm}$ [3], and the energy barrier $E = 1.5\,\mathrm{eV}$ [6].

$$k_{\mathrm{rec}} = 4\pi a_r (D_I + D_V) \exp\left[-\frac{E}{kT}\right]. \tag{3}$$

2.3. Thermal stresses

The thermal stresses in the crystal are calculated in an axisymmetric manner as described in [7]. The physical parameters are: Poisson's ratio 0.25, Young's modulus $1.56 \cdot 10^{11}\,\mathrm{Pa}$, coefficient of thermal expansion of crystal $3.2 \cdot 10^{-6}\,\mathrm{K}^{-1}$. The influence of the thermal stresses on the formation enthalpy of point defects is considered using the parameters for the isotropic case from [8], given by $\Delta H_I^f = -0.068\sigma_{\mathrm{ave}}$, $\Delta H_V^f = +0.153\sigma_{\mathrm{ave}}$, where enthalpy is expressed in eV, and stress – in GPa. The mean thermal stress $\sigma_{\mathrm{ave}} = (\sigma_{rr} + \sigma_{zz} + \sigma_{\phi\phi})/3$.

3. Results and discussion

Cone and cylindrical growth (crystal diameter 200 mm) stages are considered in the same way as it was previously done in [9]. Additionally, a crystal pull rate drop is applied following the experiment by Abe, see [5, 10] and references therein: during the cylindrical phase V_{crys} is changed from 1 mm/min to 0.2 mm/min for 30 minutes. The hot zone geometry (and hence, the heater power) is different from the one used in [10].

As noted in [10], a heater power jump should be applied in advance to counteract the large crystal diameter changes due to V_{crys} jump. This turned out to be difficult to achieve in our calculations, contrary to models where the diameter is fixed [5, 10], therefore several power curves have been tried. The considered velocity and power curves are plotted in Fig. 2, where V0 and P0 denote the reference case with no jumps. V1 corresponds to the experiment by Abe [10], while V2 is a test case to investigate the long-term pull rate change. In the case P1 the power jump of +5 kW is applied 40 min before the pull rate jump; for "P control" the heater power PID controller is used to stabilize the diameter (the shown curve was obtained after a number of repeated calculations, similarly as in [9]). It should be noted that in the calculations [5] no power change is mentioned, and the crystal radius seems to stay perfectly constant.

The crystal radius change in the cases (V1, P0), (V2, P0) and (V1, P1) exceeds 25 mm, see Fig. 3a. The power control works very well, keeping the diameter deviation below 6 mm.

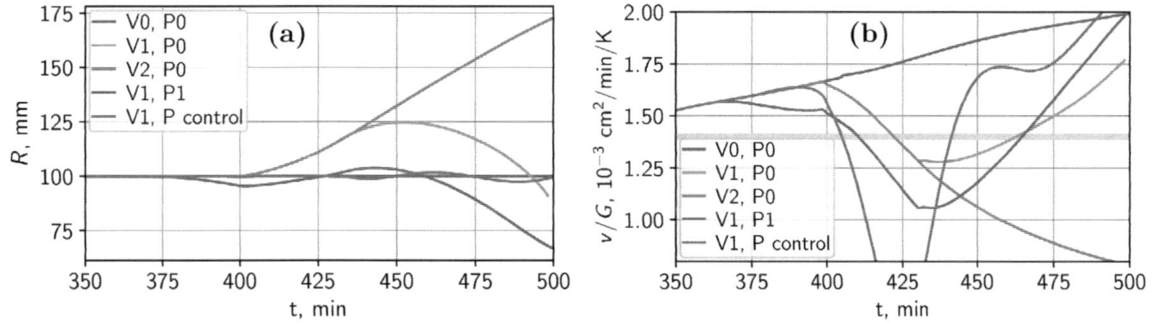

Figure 3. Time-dependence of (a) crystal radius and (b) v/G ratio at centre.

IOP Conf. Series: Materials Science and Engineering **355** (2018) 012003 doi:10.1088/1757-899X/355/1/012003

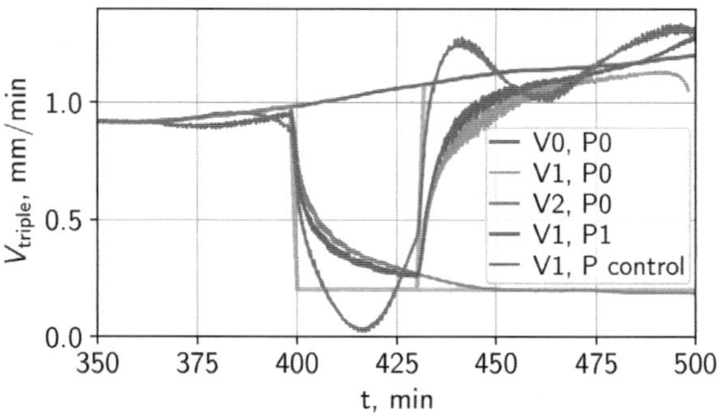

Figure 4. Time-dependence of the growth rate at the triple point V_{triple}. Solid grey line – crystal pull rate V_{crys} (see also Fig. 2a).

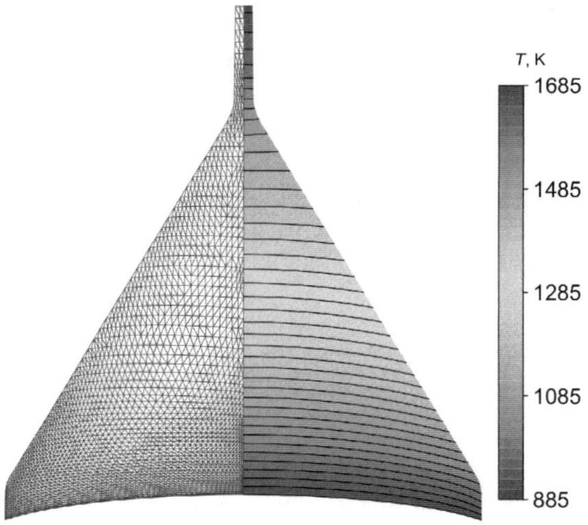

Figure 5. The finite element mesh and temperature field in the crystal. Case (V0, P0), $t = 5.5\,$h, spacing between isotherms 20 K.

According to Voronkov [11], the dominating point defect specie is determined by the ratio $\Gamma = v/G$, where v is the crystal growth rate and G – the temperature gradient. Above the critical value $\Gamma_{\text{crit}} \approx 1.4 \cdot 10^{-3}\,\text{cm}^2/\text{min}/\text{K}$ the crystal will be vacancy-rich, and below – interstitial-rich. The calculation results shown in Fig. 3b (thick grey line – Γ_{crit}), indicate that the pull rate jump indeed can introduce interstitials in the originally vacancy-rich crystal.

The response of the growth rate at the triple point on the crystal pull rate and power jumps is shown in Fig. 4; the crystal pull rate is also depicted (solid grey line). An exponential time dependence is observed. During the jump the growth rate does not catch up with the pull rate, the case with a prolonged jump (V2) shows that 50 min are needed for it to happen.

A fine mesh in the crystal has to be used for the thermal stress and point defect calculations. An example of the quasi-structured grid and the calculated temperature field is shown in Fig. 5.

Fig. 6 shows the calculated thermal stresses in the crystal. Near the crystallization interface the mean stress $\sigma_{\text{ave}} < 0$ close to the centre, and $\sigma_{\text{ave}} > 0$ close to the crystal rim. According to subsection 2.3 and [8, 12], more vacancies will be formed at the centre, and more interstitials – at the rim. Therefore, the thermal stresses increase the radial inhomogeneity of the point defect

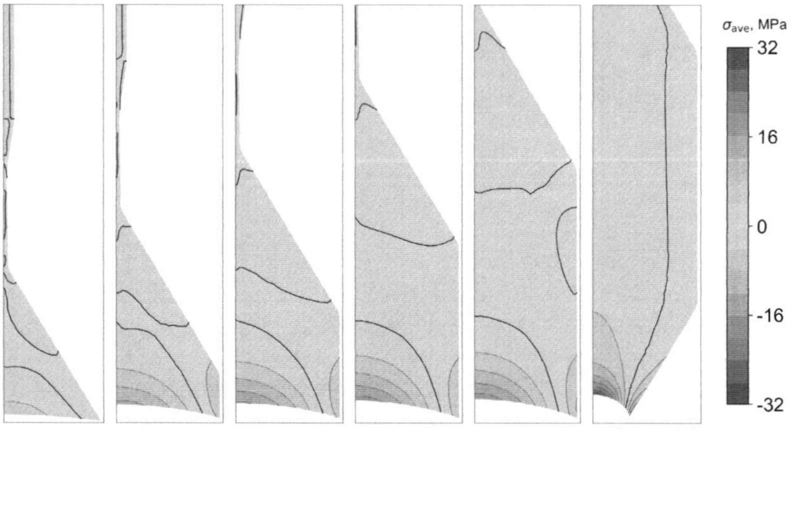

Figure 6. Mean thermal stress σ_{ave} in the crystal for the case (V0, P0) at $t = 5, 6, 7, 8, 9, 10\,$h (from left to right).

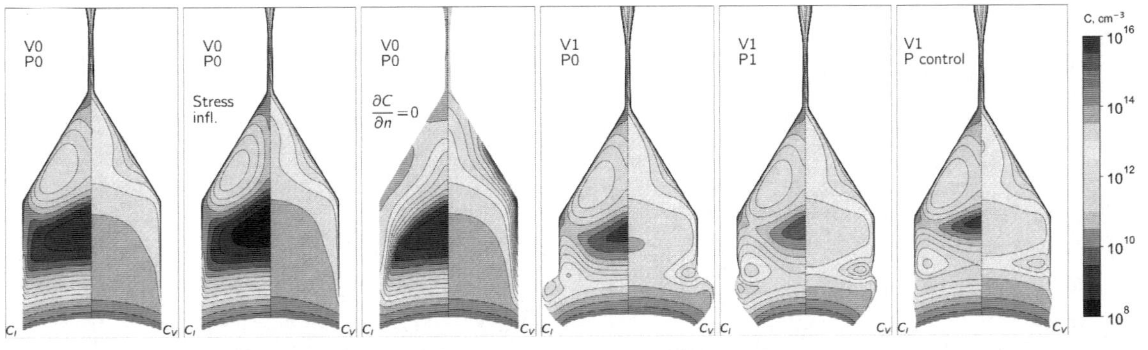

Figure 7. Point defect concentrations in the crystal at $t = 8.28\,\text{h}$.

distribution (see Figs. 7-8 "V0 P0" vs. "V0 P0, Stress infl.").

The calculated point defect distributions in the crystal at the same time instant are shown in Fig. 7; "Stress infl." means that the influence of the thermal stresses on the point defects has been considered, and "$\partial C/\partial n = 0$" – that the no-flux BC has been set at the crystal side surface. The thermal stresses increase C_V at the centre, as expected, but the overall influence is weak. The no-flux BC only slightly changes the concentrations near the crystallization interface, and considerably – further away from it. The cases with velocity curve V1 and various power curves have the same qualitative behaviour: before the radius increase a local C_I minima and C_V maxima develops near the crystal side surface, and when the radius starts to decrease, an additional C_I maxima develops.

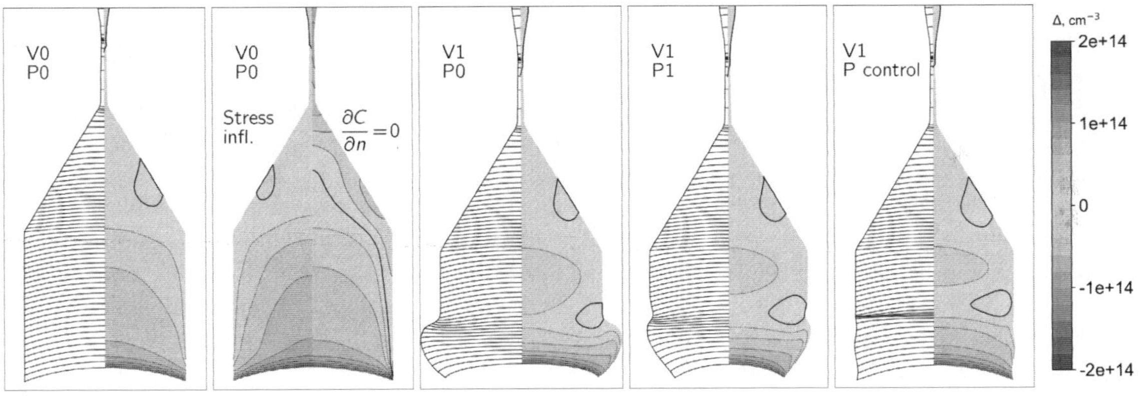

Figure 8. Crystallization interface shape plotted each 200 s (left) and Δ field in the crystal at $t = 8.28\,\text{h}$ (right). Thick solid line – zero Δ isoline.

The crystallization interface shape plotted in Fig. 8, left, visualizes the slow growth rate at the triple point and indicates that the decrease in the growth rate for the points near the centre is not as strong. The very low growth rate at the crystal rim for the power control case is clearly visible, see also Fig. 4.

According to the Δ field (Fig. 8, right), in the basic case the crystal is vacancy-rich, except for a small region in the cone. The thermal stresses make it slightly more vacancy-rich, while the no-flux BC make the outer region of the cylindrical part and the cone interstitial-rich. As observed before for the concentrations, Δ field for the velocity curve V1 and different power curves is qualitatively the same: an interstitial-rich region develops at the crystal surface due to the crystal pull rate drop.

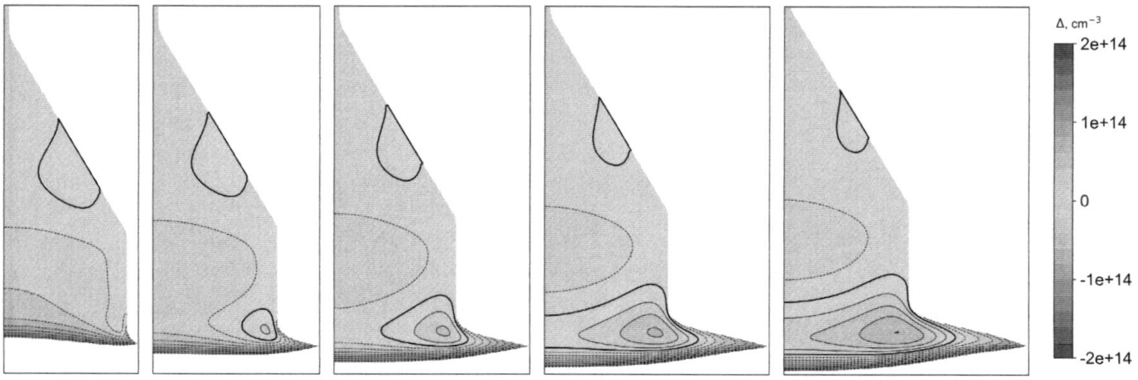

Figure 9. Δ field in the crystal for the case (V2, P0) at t = 7, 7.5 8, 8.5, 9 h (from left to right). Thick solid line – zero Δ isoline.

The I-V boundary does not reach the symmetry axis, which agrees with [5, 10]. However, it is known from these sources that when applying the same V_{crys} jump for crystals with smaller diameter, the I-V boundary reaches the centre. Since we modelled only one crystal (diameter 200 mm), it was decided to apply the pull rate drop indefinitely, case (V2, P0). The results are shown in Fig. 9: the crystal diameter changes by more than 140 mm, and, as a result of the considerably reduced v/G ratio (Fig. 3b), the diffusive flux of self-interstitials becomes dominating and leads to the increase of C_I and Δ above the crystallization interface. About 1.5 h after the jump the I-V boundary spreads to the symmetry axis, and the height of the interstitial-rich region continues to increase over time.

4. Conclusions

Non-stationary numerical modelling of CZ silicon single crystal growth has been carried out, including calculation of thermal stresses and point defects in the crystal. The influence of the thermal stresses on the point defects was tested and found out to be weak.

A qualitative reproduction of the pull rate drop experiment was achieved. Due to the large change in the crystal diameter, a power jump was applied, but, in contradiction with the experiment, the diameter changes still were considerable. Instead, a rather non-trivial power curve obtained by a PID controller was very effective in keeping the diameter constant. A longer period of a reduced pull rate was needed for the I-V boundary to reach the centre of the considered $D = 200$ mm crystal.

References

[1] Duffar T (ed) 2010 *Crystal Growth Processes Based on Capillarity* (Chichester: John Wiley & Sons)
[2] Sabanskis A, Bergfelds K, Muiznieks A, Schröck T and Krauze A 2013 *J. Cryst. Growth* **377** 9–16
[3] Sinno T, Brown R A, von Ammon W and Dornberger E 1998 *J. Electrochem. Soc.* **145** 302–18
[4] Sinno T 2007 *J. Cryst. Growth* **303** 5–11
[5] Dornberger E, von Ammon W, Virbulis J, Hanna B and Sinno T 2001 *J. Cryst. Growth* **230** 291–99
[6] Prostomolotov A I and Verezub N A 2001 *Proc. Int. Conf. on Solid State Crystals 2000* pp 97–103
[7] Muižnieks A, Raming G, Mühlbauer A, Virbulis J, Hanna B and Ammon W 2001 *J. Cryst. Growth* **230** 305–13
[8] Sueoka K, Kamiyama E, Vanhellemont J and Nakamura K 2014 *Phys. Status Solidi B* **251** 2159–68
[9] Sabanskis A, Bergfelds K and Virbulis J 2014 *Proc. Int. Sci. Coll. Modelling for Electromagnetic Processing* pp 87–92
[10] Abe T and Takahashi T 2011 *J. Cryst. Growth* **334** 16–36
[11] Voronkov V V and Falster R 1999 *J. Appl. Phys.* **86** 5975–82
[12] Sattler A, von Ammon W, Weber M, Haeckl W and Schmidt H 2007 US patent 8,043,427

IOP Conf. Series: Materials Science and Engineering **355** (2018) 012004 doi:10.1088/1757-899X/355/1/012004

Validation of mathematical model for CZ process using small-scale laboratory crystal growth furnace

Kristaps Bergfelds, Andrejs Sabanskis and Janis Virbulis

Faculty of Physics and Mathematics, University of Latvia, Zellu Street 23, LV-1002 Riga, Latvia

E-mail: kristaps.bergfelds@lu.lv

Abstract. The present material is focused on the modelling of small-scale laboratory NaCl-RbCl crystal growth furnace. First steps towards fully transient simulations are taken in the form of stationary simulations that deal with the optimization of material properties to match the model to experimental conditions. For this purpose, simulation software primarily used for the modelling of industrial-scale silicon crystal growth process was successfully applied. Finally, transient simulations of the crystal growth are presented, giving a sufficient agreement to experimental results.

1. Introduction

Semiconductor microelectronics is a large market with sales reaching 335 billion dollars [1]. This market is based on single-crystal (especially silicon) wafer production, where the first technologically complex process is the single crystal growth. Additionally, 40% of solar panels are manufactured from monocrystalline silicon (Si) [2]. Si single crystals are mostly produced by utilizing the Czochralski (CZ) and floating zone (FZ) methods [3]. Continuous crystal growth process development is necessary mostly because of the need to decrease the size of microelectronic circuitry elements [4] and increase the wafer diameter [5]. This leads to demand for larger crystals with higher quality. Larger amounts of polycrystalline Si and larger crucibles used for each crystal growth process causes large financial losses due to any faults during this process and makes large scale industrial experiments extremely expensive. Therefore, mathematical modelling is a key instrument used for the development of the crystal growth processes.

Mathematical modelling makes it possible to obtain information about important physical parameters relevant to the growth process without performing costly experiments. The authors' research group at the University of Latvia (UL) Faculty of Physics and Mathematics has developed such mathematical modelling tools for the simulation of the CZ crystal growth process [6]. However, it is necessary to perform validation of the mathematical model to ensure its relation to an actual crystal growth process by using practical experiments. As mentioned before, industrial scale experiments are too expensive to be considered as routine tool for mathematical model development. In addition to large-scale experiments, small laboratory scale research techniques must be established to perform cost-effective validation of the developed mathematical methods.

Content from this work may be used under the terms of the Creative Commons Attribution 3.0 licence. Any further distribution of this work must maintain attribution to the author(s) and the title of the work, journal citation and DOI.
Published under licence by IOP Publishing Ltd

2. Problem definition

Authors of this article have access to small-scale laboratory CZ crystal growth furnace (figures 1 and 2) which can be used for the mathematical model verification described previously. This furnace consists of an electrically heated crucible unit (see schematics in figure 3) and a crystal puller. Crucible contains about 150 ml of NaCl-RbCl salt mixture with molar ratio of 40:60. This composition has eutectic properties which allow for relatively low melting point around 815 K [7, 8]. Optimal growth conditions are achieved at pulling speeds of 0.6 mm/min (see illustrative photography in figure 1). Crystals can be grown with diameter around 20 mm (see example in figure 2). After performing experimental crystal growth shown in figure 2, it is possible to study the application of the available mathematical modelling techniques for the simulation of this experiment — which is the main topic of this article.

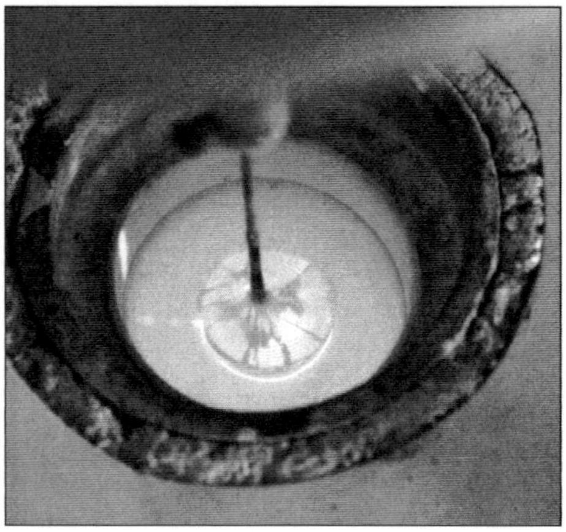

Figure 1. Photography of the NaCl-RbCl salt crystal growth process.

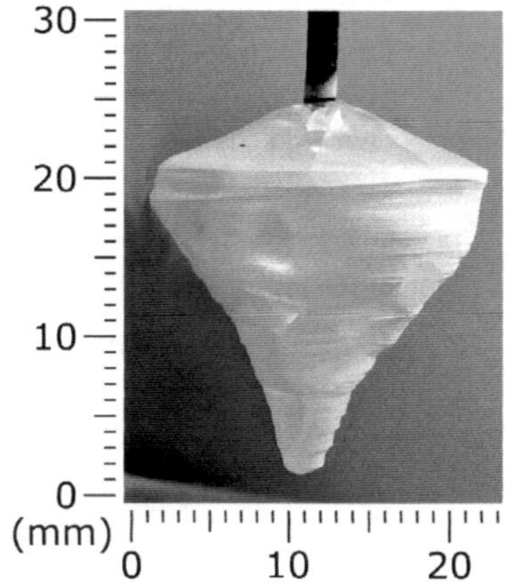

Figure 2. Example of a grown crystal.

Although the crystal growth performed with this material first seems as considerably different from the industrial CZ-Si growth processes (size, temperature range, presence of atmosphere, translucency), system still exhibits all the physical phenomena that can be related to silicon crystal growth and does not have the drawbacks associated with some liquid metals e.g. gallium and its eutectics. Namely, the crystal growth can be performed in room conditions and does not require a shielding liquid or a pressure vessel to protect it from air. Additionally, the temperature field structure in our furnace is similar to industrial CZ process, contrary to gallium which melts at 30°C and might require cooling — instead of heating — to be used in crystal growth experiments.

3. Calibration of model parameters

3.1. Overview of the mathematical model and creation of geometry

For the studies performed in this article, previously developed simulation software "CZ-Trans" was used [6]. It can perform axisymmetric non-stationary simulations of CZ crystal growth process. The following physical phenomena are taken into account: (1) heat transfer by conduction, (2) heat transfer by thermal radiation (view factor approach as in [9]), (3) shape of the melt free surface, (4) shape of crystallization interface and (5) shape of crystal side surface.

To perform simulations with the software, axisymmetric geometry was created by simplifying various system parts to speed up simulations and conform to the limitations of the model (figure 4). As it can be seen, ceramic parts are unified into a single domain and separate

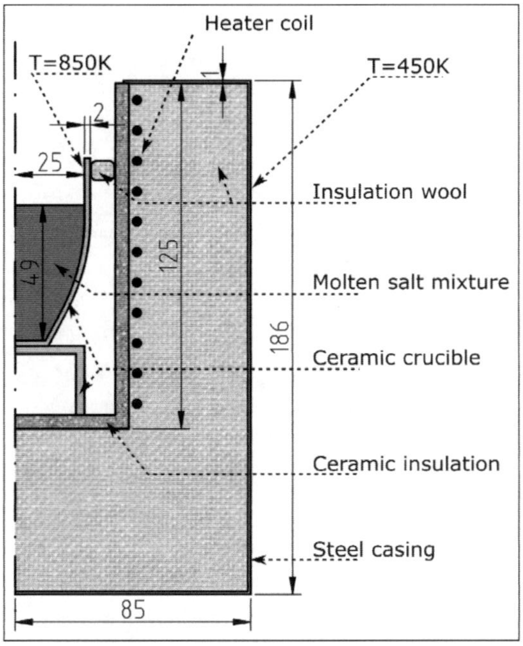

Figure 3. Precise geometry of the crucible-heater unit used in the laboratory crystal growth furnace (dimensions in mm).

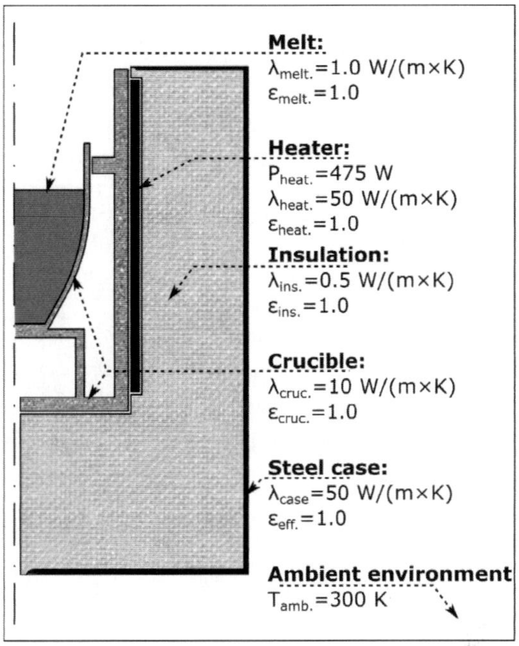

Figure 4. Simplified geometry for the creation of FEM domains used in numerical simulations.

windings of the heater wire are modelled by a single FEM block. These simplifications are required to correctly model the radiation dominating heat transfer between these components.

3.2. Data from experimental heating of crucible as means for model calibration

During the experiment, before the crystal growth shown in figure 2 could be performed, crucible was heated up using its full electrical power of 475 W until all its contents were melted and stationary temperature field was achieved. This situation was characterized by the following temperature measurements using a pyrometer: 450±10 K on the outer steel casing and 850±20 K on the crucible edge, just besides melt free surface. This information can be used to calibrate the parameters used in the software to achieve reasonably realistic temperature distribution for numerical simulations.

Physical parameters for the melt [7] and corundum ceramic parts [10] are well known (see figure 4 for summary). However, the thermal conductivity of the silica insulation wool used in the construction of the crucible is not known. This parameter ($\lambda_{\text{ins.}}$) can vary from 0.05 to 0.6 W/(m·K) depending on temperature and how tightly it is packed [11]. Because of the highly empirical nature of this parameter, it was chosen as means of calibration of our model — by varying this parameter we would ensure that the calculated temperature field corresponds to the measurements done with pyrometer.

3.3. Approach to convective heat transfer modelling

"CZ-Trans" software currently does not support 3rd type boundary condition and, therefore, convective heat transfer from steel casing to ambient air require a specialized modelling approach. To achieve realistic temperature field within the system, emissivity of the steel case was increased as means to introduce additional heat flux through the insulation layer.

The approach to introduce effective emissivity $\varepsilon_{\text{eff.}}$ instead of using the actual material property $\varepsilon_{\text{surf.}}$ can be related to the use of effective heat transfer coefficient $h_{\text{surf.}}$ described in [9]. First, we assume that the heat flux from the steel casing of the crucible block consists of two parts: radiative and convective heat losses (equation 1).

$$q_{\text{surf.}} = q_{\text{rad.}} + q_{\text{conv.}} \tag{1}$$

Then we can write these heat fluxes explicitly by using the effective emissivity $\varepsilon_{\text{eff.}}$ and heat transfer coefficient $h_{\text{surf.}}$ (equation 2).

$$\varepsilon_{\text{eff.}}.\sigma_{\text{SB}}(T^4_{\text{surf.}} - T^4_{\text{amb.}}) = \varepsilon_{\text{surf.}}.\sigma_{\text{SB}}(T^4_{\text{surf.}} - T^4_{\text{amb.}}) + h_{\text{surf.}}.(T_{\text{surf.}} - T_{\text{amb.}}) \tag{2}$$

We then solve the equation 2 for $h_{\text{surf.}}$ to evaluate how large is the heat transfer coefficient effectively applied in our simulation, when we find the appropriate $\varepsilon_{\text{eff.}}$ for the particular set of parameters and temperatures (equation 3).

$$h_{\text{surf.}} = \frac{\sigma_{\text{SB}}(T^4_{\text{surf.}} - T^4_{\text{amb.}})(\varepsilon_{\text{eff.}} - \varepsilon_{\text{surf.}})}{T_{\text{surf.}} - T_{\text{amb.}}} \tag{3}$$

It must be noted that using this approach (applying effective emissivity instead of 3rd type boundary condition) is entirely suitable with $\varepsilon_{\text{eff.}} > 1$ as it is only means to achieve the required heat loss from a surface that radiates only to ambient environment.

3.4. Simulation results of a stationary temperature field matched to the experiment

At this point during the modelling process, there are two unknown parameters that need to be adjusted to achieve realistic temperature field for the FEM simulations — insulation thermal conductivity $\lambda_{\text{ins.}}$ and effective emissivity of the steel casing $\varepsilon_{\text{eff.}}$ — which is the necessary amount of free parameters to match the two experimentally measured temperatures: of the crucible edge and outer steel casing (see table 1). A series of calculations were performed to achieve this goal.

The final (fitted) calculated temperature field as well as simulation FEM mesh is shown in figures 5 and 6. We can see that the temperatures are indeed matched to the measured values mentioned earlier — calculated values range from 450 to 550 K on the outer steel casing and from 850 to 900 K on the crucible edge.

The summary of the optimized parameters and their relationship to values found in literature are shown in table 1. The optimal emissivity $\varepsilon_{\text{eff.}}$ that ensures this temperature field is 1.0. The heat exchange coefficient $h_{\text{surf.}}$ that corresponds to such a choice is 7.5 $\mathrm{W/(m^2{\cdot}K)}$ which is entirely realistic given the data in [9]. It can be concluded that the main thermal parameters of our model are successfully fitted to perform transient simulations of the crystal growth experiment.

4. Transient simulations

4.1. Simulation of the of the laboratory crystal growth experiment

After performing the model calibration, it was possible to use our transient simulation capabilities to replicate the experimental growth conditions as close as possible. PID control algorithms were used to optimize the growth parameters (pull rate and heater power) to match the experimental crystal shape. As seen in figures 7 and 8, calculated crystal shape is matched considerably well to the experimental shape seen in figure 2. It must be noted that temperature field and heater power values are also consistent with experimental conditions.

Despite the reasonable precision of the crystal shape modelling, figure 7 shows that the desired pull rate of 0.6 $\mathrm{mm/min}$ during the simulation is not maintained. This fact is due to the way PID controller cascade is set up: 1) pull rate is adjusted to maintain the required shape and 2) the heater power is adjusted to minimize the pull rate adjustment.

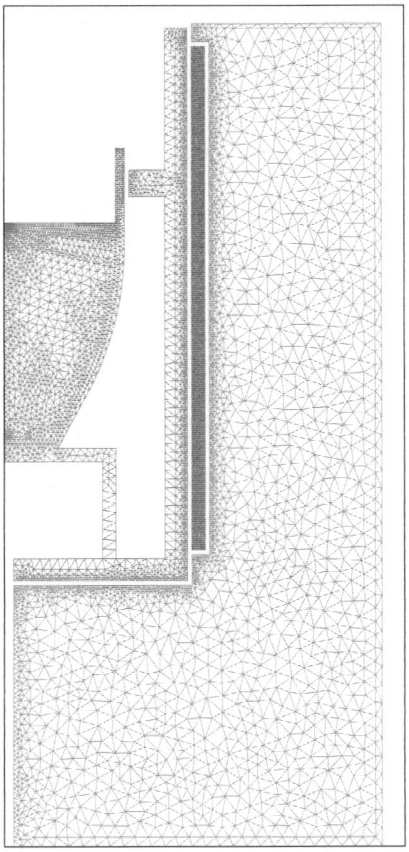

Figure 5. Mesh used for FEM simulations of temperature field within the crucible-heater unit.

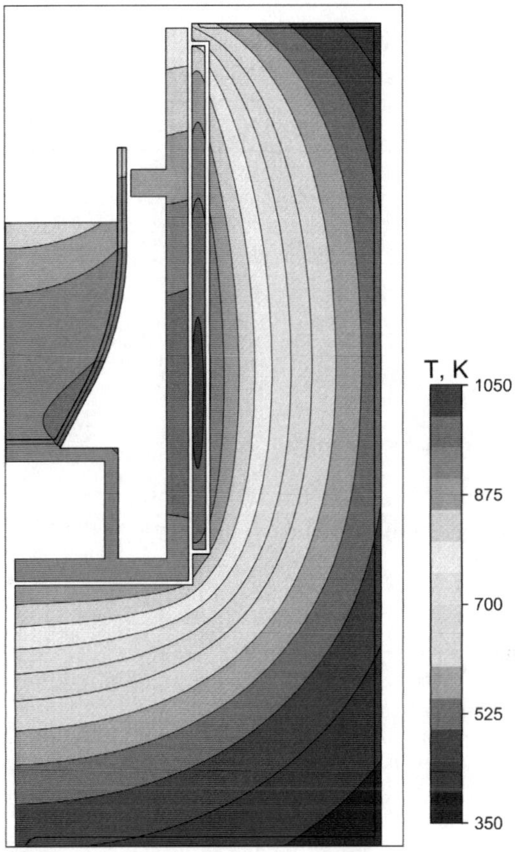

Figure 6. Result of the stationary temperature field calculation.

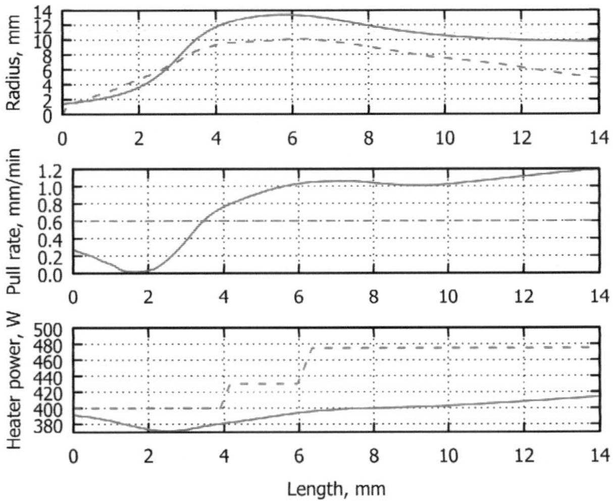

Figure 7. Crystal radius, pull rate and heater power during the laboratory crystal growth experiment (dashed line) and transient simulation (solid line).

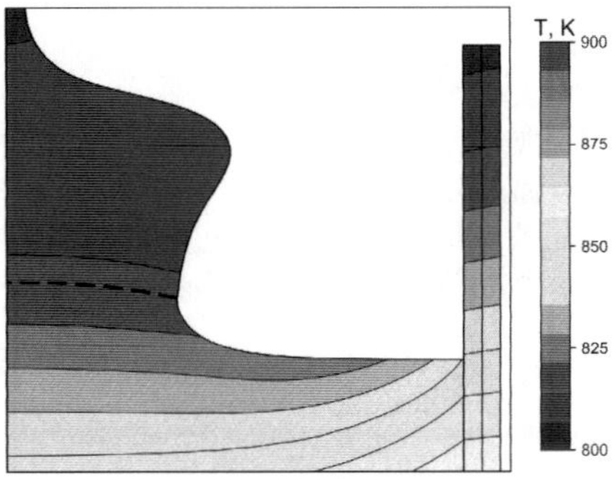

Figure 8. Temperature field and phase boundaries at the end of transient simulation.

Table 1. Summary of known and calibrated parameters that are featured in the initial stationary temperature field simulations.

Property	**Denotation**	**Value**	**Notes**
Ambient temperature	$T_{\mathrm{amb.}}$	300 K	Experiment was performed in room temperature.
Steel casing covered with aluminium paint	$\varepsilon_{\mathrm{surf.}}$	0.4	Value found in [9]. Value increased in simulation to account for convective cooling (see $\varepsilon_{\mathrm{eff.}}$ below).
Average crucible edge temperature	T_{edge}	850 K	Measured experimentally, **must be matched by simulation**.
Average steel casing temperature	$T_{\mathrm{surf.}}$	450 K	Measured experimentally, **must be matched by simulation**.
Effective emissivity of steel casing	$\varepsilon_{\mathrm{eff.}}$	1.0	Free **parameter that was optimized** during numerical studies to match temperature field to measured values.
Heat exchange coefficient for steel casing	$h_{\mathrm{surf.}}$	7.5 $\mathrm{W}/(\mathrm{m}^2{\cdot}\mathrm{K})$	Value that corresponds to $\varepsilon_{\mathrm{eff.}}$ listed above. This corresponds to typical values for free convection found in [9].
Insulation thermal conductivity	$\lambda_{\mathrm{ins.}}$	0.5 $\mathrm{W}/(\mathrm{mK})$	Free **parameter that was optimized** during numerical studies to match temperature field to measured values.

Evidently, the experimental heater power curve is not suitable for the modelling case and, therefore, large deviation of the pull rate is necessary to maintain the crystal shape. This problem can be solved by iteratively applying the PID-produced power curve for consecutive simulation cases and shifting it towards earlier times. In this way, the temperature near the crystal will be exactly right, despite thermal inertia of the whole furnace.

Furthermore, figure 8 shows a concave crystallization interface. This is not consistent with experimental conditions where it has been observed (by abruptly pulling out the crystal from the melt) that the interface is convex — central part is located lower in the melt.

4.2. Simplified model for heat losses through the transparent crystal

The difference of crystallization interface shape between the model and experimental conditions is related to the transparency of the crystal. Presented results consider mathematical model where the radiative heat transfer within the melt and crystal is not included. However, we propose a simplified model that can ensures sufficient precision of simulated phase boundary shape, but do not require substantial model development effort. By knowing the temperature (melting point) and emissivity of the crystallization interface [12] the corresponding radiative heat flux can be calculated as shown in equation 4.

$$q_{\text{rad.}} = \varepsilon\sigma_{\text{SB}}T^4 = 0.962 \cdot 6.67 \cdot 10^{-8} \cdot 815^4 = 24.1\,\text{kW}/\text{m}^2 \quad (4)$$

The calculated heat flux density is assumed to be lost (radiated) from the crystallization interface and transferred away without absorption and reflection. Therefore, it was applied as a constant negative heat source on the crystallization boundary in our model. By using this approach, results seen in figure 10 are achieved — the crystallization interface is indeed concave, giving better match to experimental conditions.

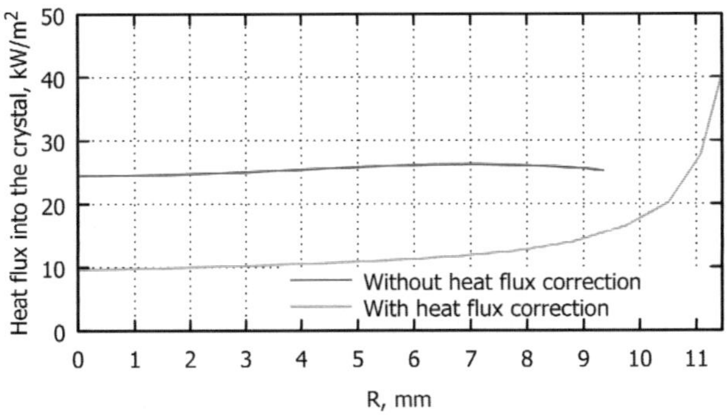

Figure 9. Heat flux density distribution on the crystallization interface for the both transient simulations presented in this article — with and without heat flux correction.

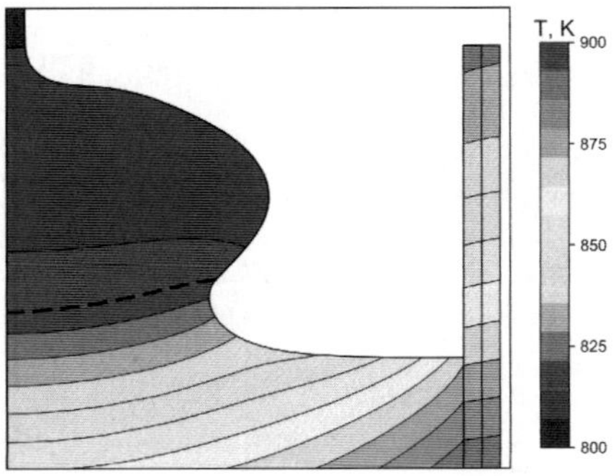

Figure 10. Temperature field and phase boundaries at the end of transient simulation (with applied correction of melt-crystal interface shape).

By introducing the negative heat sources on the interface, the total conductive heat flux from the interface into the crystal is reduced significantly (figure 9) and the crystallization occur more rapidly. As a result, the presented cases (figures 8 and 10) with roughly the same crystal length of 15 mm have considerably different duration — 28 minutes without the heat flux correction and only 12 minutes for the corrected case. Additionally, crystal has larger diameter as the predefined (experimental) crystal shape is increasingly harder to maintain by using the PID algorithms with the same parameters.

5. Conclusions

Calculation software used primarily for the modelling of industrial-scale silicon crystal growth process has been successfully used for the optimization of unknown material properties of a small-scale laboratory NaCl-RbCl crystal growth furnace. Stationary temperature field that corresponds to experimentally measured temperatures have been calculated. This result was obtained by choosing thermal conductivity of $0.5\,\text{W}/(\text{m}\cdot\text{K})$ for the thermal insulation layer and choosing effective emissivity coefficient of 1.0 for the outer steel case. Such approach is equivalent to using convective heat transfer coefficient of $7.5\,\text{W}/(\text{m}^2\cdot\text{K})$.

The obtained stationary calculation results and model parameters were used as basis for the transient simulation of the laboratory crystal growth experiment. Sufficient match between the modelled and experimental crystal shape was achieved, although additional studies are necessary to acquire optimal heater power distribution that ensures the required pull rate of $0.6\,\text{mm}/\text{min}$.

Simplified model of heat losses from crystallization interface due to radiative heat transfer through the transparent crystal was proposed. By introducing additional, constant heat flux density on the crystallization interface, a convex crystallization interface shape was achieved — a result that corresponds to experimental conditions.

Acknowledgments
The research was carried out with the financial support from SIA "Mikrotīkls". University of Latvia Foundation project *"Implementation of Experimental Laboratory Crystal Growth Furnace"* (project number: 2183).

References
[1] Statistics Portal 2016 *Global semiconductor sales*
[2] Results Revision 2014 *Int. Technology Roadmap for Photovoltaic*
[3] Duffar T 2010 *Crystal Growth Processes Based on Capillarity* (Wiley-Blackwell)
[4] Collaert N 2014 *Microelectronic Eng.* **132** 218-25
[5] SEMI.ORG 2015 *Macroeconomics of 450mm Wafers*
[6] Sabanskis A, Bergfelds K, Muiznieks A, Schröck T and Krauze A 2013 *J. Crystal Growth* **377** 9-16
[7] Centre for Research in Computational Thermochemistry 2017 *FACT salt database*
[8] Janz G J 1967 *Molten Salts Handbook* (Academic Press)
[9] Incropera F P and DeWitt D P 1996 *Introduction to Heat Transfer* (Wiley)
[10] Lide D R 1998 *Handbook of Chemistry and Physics* (CRC Press)
[11] Heraeus Quarzglas 2017 *Quartz wool technical specification*
[12] International Crystal Laboratories 2017 *NaCl Optical Crystals*

IOP Conf. Series: Materials Science and Engineering **355** (2018) 012005 doi:10.1088/1757-899X/355/1/012005

Three-dimensional modelling of thermal stress in floating zone silicon crystal growth

Matīss Plāte[1], Armands Krauze[2] and Jānis Virbulis[2]

[1] Siltronic AG, Hanns-Seidel-Platz 4, 81737 München, Germany

[2] University of Latvia, Faculty of physics and mathematics, Zeļļu 23, LV-1002, Latvia

E-mail: matiss.plate@siltronic.com

Abstract.
During the growth of large diameter silicon single crystals with the industrial floating zone method, undesirable level of thermal stress in the crystal is easily reached due to the inhomogeneous expansion as the crystal cools down. Shapes of the phase boundaries, temperature field and elastic material properties determine the thermal stress distribution in the solid mono crystalline silicon during cylindrical growth. Excessive stress can lead to fracture, generation of dislocations and altered distribution of intrinsic point defects. Although appearance of ridges on the crystal surface is the decisive factor of a dislocation-free growth, the influence of these ridges on the stress field is not completely clear. Here we present the results of thermal stress analysis for 4" and 5" diameter crystals using a quasi-stationary three dimensional mathematical model including the material anisotropy and the presence of experimentally observed ridges which cannot be addressed with axis-symmetric models. The ridge has a local but relatively strong influence on thermal stress therefore its relation to the origin of fracture is hypothesized. In addition, thermal stresses at the crystal rim are found to increase for a particular position of the crystal radiation reflector.

1. Introduction

The industrial floating zone (FZ) method for the growth of silicon single crystals provides semiconductor material of lowest impurity concentration which is mainly used for high power devices. The main parts of the process inside the growth chamber are the cylindrical poly crystalline silicon feed rod being pushed downwards, the single turn high-frequency induction coil to melt the feed rod through its needle-eye opening, the molten zone just below the inductor and the growing crystal which supports the molten zone. The crystal is axially rotated therefore cylindrically shaped ingot is obtained and crystallization interface is normally concave because the latent heat at the centre can not be removed as easily as at the periphery close to the radiating crystal surface. The non-flat crystallization interface and the thermal radiation from the surface gives a radial component of temperature gradient resulting in an inhomogeneous expansion and possibly high levels of thermal stress. The consequences of the stress with its considerable influence on the process and product quality can be divided into three categories:

- Fracture of the crystal when the stress is above some critical level. The sudden and unpredictable fracture leads to a complete material loss and a damage to the growth equipment due to the spillage of the melt. Risk of the fracture increases with crystal diameter.

Content from this work may be used under the terms of the Creative Commons Attribution 3.0 licence. Any further distribution of this work must maintain attribution to the author(s) and the title of the work, journal citation and DOI.
Published under licence by IOP Publishing Ltd

- Dislocation generation, the probability of which correlates to the stress level and also leads to a process failure in a less destructive manner. Large diameter single crystals can not be grown with dislocations [1].
- Altered incorporation of the intrinsic point defects as thermal stress influences the Voronkov's criterion, the critical ratio v/G or $\Gamma 0_{\mathrm{crit}}$. Under compressive stress the $\Gamma 0_{\mathrm{crit}}$ tends to decrease and vice versa [2].

With the needle-eye technique, the capillary limitation of a cylindrical zone is surmounted. However, currently the ingot diameter is limited to 200 mm with the two main impediments being the crystal fracture due to thermal stress with related dislocation generation and electrical breakdown at the inductor main slit [3].

One of the first reported calculations of thermal stress in the silicon crystal during FZ growth was part of a global axis-symmetric quasi-steady model including electromagnetic field and heat transfer for a 4" crystal diameter [4]. Analytic approach for the stress was used and various inductor designs were compared with the aid of the model. Maximum compressive stress at the centre of the crystallization interface and a smaller maximum of tensile stress on the crystal surface at or close to the triple point was found and the scalar von Mises stress was also used as a measure.

3D anisotropic model of thermal stress for 6" and 8" crystal diameters using the Czochralski (Cz) method predicted dislocation generation in all of the considered cases [5]. Using the knowledge of the discrepancy between predictions by existing models and experimental observations, a concept of a metastable state of the crystal where the energy necessary to generate dislocations is small compared to the total energy of the crystal has been introduced [6]. The higher total energy of the crystal the smaller is the energy barrier for dislocation generation and random perturbations can lead to dislocation generation more easily. The idea supported the results of axis-symmetric finite element simulations for 4" FZ crystals where exceed stress was used to predict dislocations in a crystal region spanning from the crystallization interface until approximately a radius below. Reduction of the thermal stresses with the introduction of a crystal radiation reflector was shown as well. Global model of FZ system which includes axis-symmetric stress field using Comsol Multiphysics has been developed at the Leibniz Institute for Crystal Growth [7]. The case study revealed that higher pull rate leads to an increase of thermal stress, and presence of the reflector decreases the stress. The above-mentioned studies all consider a perfectly round crystal in contrast to a numerical study of sinusoidal surface variations for a Cz crystal based on experimental observations [8]. Thermal stress was shown to locally increase in any of the considered configurations of the periodical surface undulations. The problem of thermal stress in the Cz crystal has been further addressed by development of a three-dimensional (3D) model including the uneven surface and material anisotropy [9]. The $\langle 1\,1\,1\rangle$ crystal growth orientation was found to be of the lowest stress level.

However, experimentally observed ridges on the crystal surface [10] which can not be studied by axis-symmetric models are the main indication for a dislocation free growth. Morphology of the ridges for FZ crystals of $\langle 1\,1\,1\rangle$ orientation up to 5" diameter has been described and a loss of particular ridge shapes when the crystal dislocates is suggested [11]. Size of the ridge has been theoretically calculated for a $\langle 1\,0\,0\rangle$ 4" Cz crystal depending strongly on the temperature gradient by the crystallization interface at the triple point [12]. The model predicts larger ridge size for a smaller temperature gradient. Formation of periodic undulations of the ridge – also called bulges – are observed for $\langle 1\,1\,1\rangle$ crystals due to the inhomogeneous growth rate depending on the crystallographic orientation [13]. The necessary undercooling of the melt in the $\langle 1\,1\,1\rangle$ direction is the slowest and will move the triple point lower thus causing the free melt surface to bend outwards and the bulge to grow. Once the bulge has started to grow, the radiation heat losses from the increased surface area will provide enough undercooling for the triple point line to rise and bulge to decrease. For $\langle 1\,0\,0\rangle$ orientation mostly steady state growth of straight ridges is observed due to the different angle of {1 1 1} planes. The period of these oscillations depends on the physical conditions at the triple point line during the process which determine the temperature gradient.

Numerical simulation of the crystal ridge was first carried out by [14] for a 4" FZ ⟨1 1 1⟩ crystal. However, the standard mesh was used without refining the ridge area which can have a considerable influence of the precision of the numerical results. The mathematical model developed and shown in [14] is used for the present calculations. The same numerical procedures apply except for mesh refinement by the ridge. Smooth crystal shape is compared to surface with a ridge for 4" and 5" crystal diameters using two different pull rates and ⟨1 0 0⟩ and ⟨1 1 1⟩ crystal orientations and an undulated feature of the ridge is also introduced as reported in [15]. The crystal radiation reflector is considered for 5" crystal with 2.5 mm/min pull rate. The given modelling approach is suitable for quasi-steady conditions which correspond to the cylindrical growth phase.

2. Numerical model

2.1. Boundary conditions

To have realistic interface shapes and temperature field in the crystal, initially a converged state of quasi-steady calculations using the program FZone [16] with inductor shape from [7] is obtained for the given crystal diameter and the pull rate. Shape of the crystallization interface is found by solving global heat transfer problem which couples the temperature field in the silicon with radiation on the surfaces using view factors, and the heat sources are taken from 3D high-frequency electromagnetic field calculation [17]. The inductor current is adjusted to provide the target zone height, i.e., the vertical distance between the triple point and rim of the melting front. With crystal radiation reflector present, temperature in the reflector is calculated using emissivity and heat conductivity of the specified material.

2.2. Mesh generation and 3D geometry

A quasi-structured mesh of second order elements is used to solve the heat transfer and linear elasticity problems. First, a unit circle is meshed symmetrically with radial layers of triangular or quadrilateral elements. Second, the converged boundary shapes from the axis-symmetric FZone calculation are rotated to obtain the initial crystal domain in 3D. Then, the 2D mesh is mapped onto the crystallization interface and successive layers of the respective prism and hexahedron elements are added along the vertical and almost parallel to the crystallization interface with the desired discretization level.

Mesh for the ridge is refined in the area spanning four quadrilateral elements in the azimuthal direction and two – in the radial direction to have twice the elements on the crystal surface covering the ridge to capture any rapid changes of the temperature and the strain field, see Fig. 1. Shape of the ridge on the triple point line is approximated by analytical formula:

$$R(\phi) = R_0 + Ae^{-\left(\frac{\phi}{\phi_0}\right)^2} \tag{1}$$

where ϕ is the azimuthal angle, R_0 is the initial crystal radius, $A = 1\,\text{mm}$ is the amplitude of the ridge and $\phi_0 = 5°$ is the effective width of the ridge.

Periodic vertical undulations are introduced to the ridge by

$$R(z) = R_0 + \frac{R - R_0}{2}\left(1 - \sin\frac{z - z_0}{l}2\pi\right) \tag{2}$$

where $l = 15\,\text{mm}$ is the period of the undulations and z_0 is vertical shift. The size parameters are taken from the crystal surface photographs in [15], see Fig. 2 for the corresponding numerical model with two ridge shapes considered.

IOP Conf. Series: Materials Science and Engineering **355** (2018) 012005 doi:10.1088/1757-899X/355/1/012005

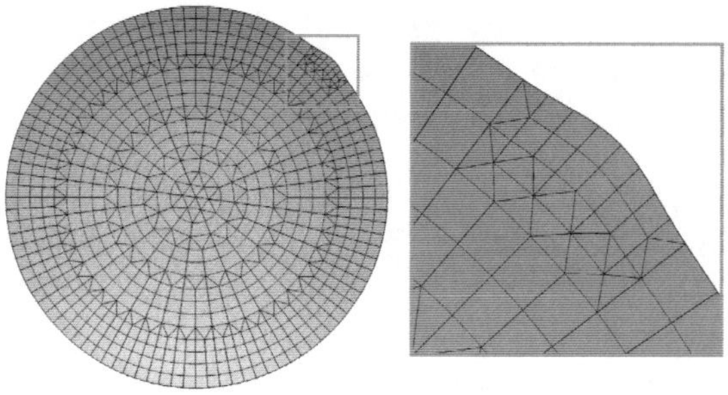

Figure 1. Quasi-structured finite element mesh on the crystallization interface with a close-up view of the refinement at the ridge.

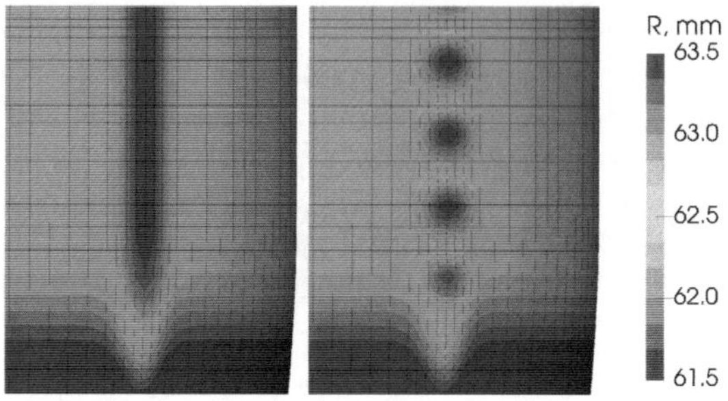

Figure 2. Surface of 5" diameter crystal with unperturbed ridge (left) and ridge with periodic undulations (right) coloured according to the radius value.

2.3. Temperature field

Heat conduction equation with temperature dependent heat conductivity λ is solved in the crystal:

$$\vec{\nabla}\left(\lambda(T)\vec{\nabla}T\right) = 0 \tag{3}$$

Constant temperature boundary condition is applied to the crystallization interface and distribution of the apparent ambient temperature for the radiation boundary condition is applied on the crystal surface. This temperature is found from the initial results by FZone using formula:

$$T_{\mathrm{amb}} = \sqrt[4]{T^4 - \frac{q_{\mathrm{rad}}}{\epsilon\sigma_{\mathrm{SB}}}} \tag{4}$$

where $T = T(z)$ is the temperature on the crystal surface, q_{rad} is the net radiation heat flux, $\epsilon = \epsilon(T)$ is the emissivity and σ_{SB} is the Stefan–Boltzmann constant.

2.4. Thermal stress field

Linear elasticity problem based on the stress-strain relationship (Hooke's law) is solved:

$$\sigma_i = C_{ij}\left(\epsilon_j - \epsilon_{Tj}\right) \tag{5}$$

where ϵ_T is isotropic thermal strain and C is the elasticity matrix. For $\langle 1\,0\,0\rangle$ crystal growth orientation, the elasticity matrix can be expressed as:

$$C = \begin{bmatrix} C_{11} & C_{12} & C_{12} & 0 & 0 & 0 \\ & C_{11} & C_{12} & 0 & 0 & 0 \\ & & C_{11} & 0 & 0 & 0 \\ & & & C_{44} & 0 & 0 \\ & \text{Sym.} & & & C_{44} & 0 \\ & & & & & C_{44} \end{bmatrix} \tag{6}$$

And the respective expression for $\langle 1\,1\,1\rangle$ orientation:

$$C = \begin{bmatrix} C_{11} - K/2 & C_{12} + K/6 & C_{12} + K/3 & -\sqrt{2}K/6 & 0 & 0 \\ & C_{11} - K/2 & C_{12} + K/3 & \sqrt{2}K/6 & 0 & 0 \\ & & C_{11} - 2K/3 & 0 & 0 & 0 \\ & & & C_{44} + K/3 & 0 & 0 \\ & \text{Sym.} & & & C_{44} + K/3 & -\sqrt{2}K/6 \\ & & & & & C_{44} + K/6 \end{bmatrix} \tag{7}$$

where $K = C_{11} - C_{12} - 2C_{44}$. Experimental measurements of the Young's modulus are available [18] and the corresponding temperature-dependence of the elastic constants has been implemented. The final equation for the problem is the force balance:

$$\frac{\partial \sigma_{ij}}{\partial x_j} = 0. \tag{8}$$

3. Calculation results

Comparison of the FZone results between 5" crystal diameter cases with and without reflector in Fig. 3 show the shapes of phase boundaries and the vertical distribution of the apparent ambient temperature on the crystal surface calculated from equation 4. Depth of the crystallization interface increases using the reflector because of the significantly higher apparent ambient temperature and therefore lower heat losses from the crystal surface. The physical parameters of the heat transfer are temperature of the ambient space 600 K, temperature of the inductor 400 K and emissivities of the inductor and the reflector 0.3 . Average temperature of the reflector 1150 K is obtained without any cooling and is close to estimated 1000 K for a realistic FZ process [16]. However, water cooling would be preferable in order to avoid deformation of the silver reflector near its melting point. With larger depth of the crystallization interface and therefore smaller interface angle, in general an increase of thermal stress is expected.

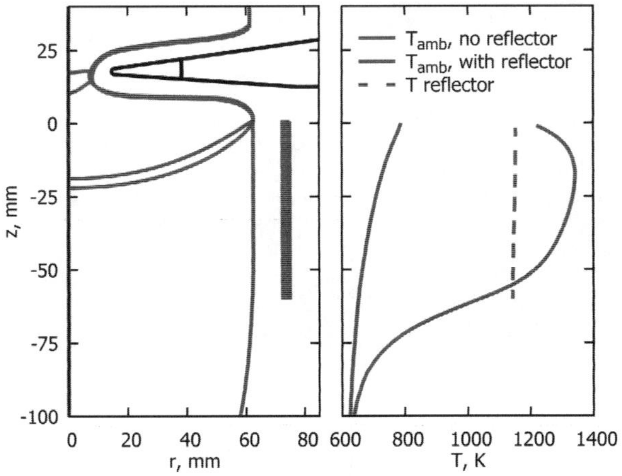

Figure 3. Comparison between FZone results of axisymmetric heat transfer for cases of 5" crystal diameter without reflector (blue lines) and with reflector (red lines). Shapes of phase boundaries with inductor given in black (left) and the apparent ambient temperature on the crystal surface and temperature of the reflector (right).

Von Mises stress invariant is commonly used as a universal measure for the stress impact on dislocation generation and fracture [3]. Influence of the ridge and process parameters on von Mises stress on the crystal surface is shown in Fig. 4 where two values of pull rate, two crystal orientations and two radii are compared. For all the cases stresses in 5" crystal are higher that in 4" crystal by roughly 20 MPa. The azimuthal variation of stress due to the ridge is considerably stronger than due to the material anisotropy. Stress maximum without the ridge can be located on the triple point line as in cases (c, d) with high pull rate and below the triple point line as in cases (a, b) that have lower pull rate. There is a sudden decrease of the stress on the ridge which is always accompanied by a pronounced stress maximum just by the side of the ridge. The maximum is located mostly up to 5 mm below the triple point line (a, c, d) but can reach 20 mm as in (b). Stress on the triple point line is approximately 10 MPa lower than the maximum by the side of the ridge below the triple point line. Larger azimuthal variations of von Mises stress along the triple point line are observed for case with the reflector as shown in Fig. 5. Stresses for a round $\langle 1\,0\,0 \rangle$ crystal have a four-fold symmetry with relatively low variations compared to $\langle 1\,1\,1 \rangle$ orientation with a more pronounced three-fold symmetry of von Mises stress. Slight increase of the stress by the side of the ridge is noticed for all cases except for the undulated ridge which gives largest stress at the tip of the ridge. This peak exceeds the maximal value for case without ridge by 1 MPa. For ⟨1 1 1⟩ crystals the stress minimum at the ridge is slightly smaller than for ⟨1 0 0⟩ crystals although the stress is larger elsewhere.

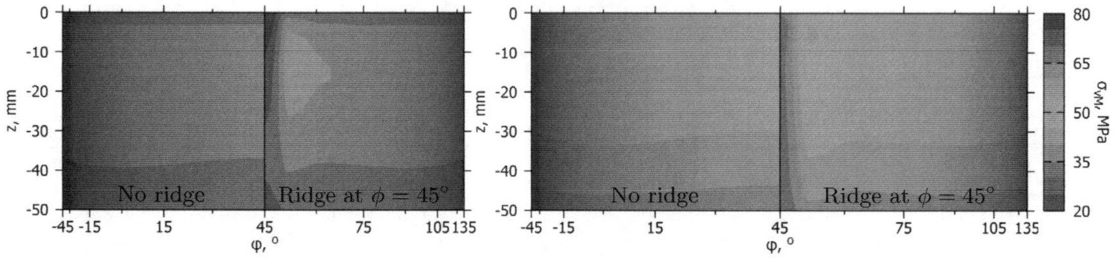

(a) ⟨1 0 0⟩, $v = 2.5$ mm/min

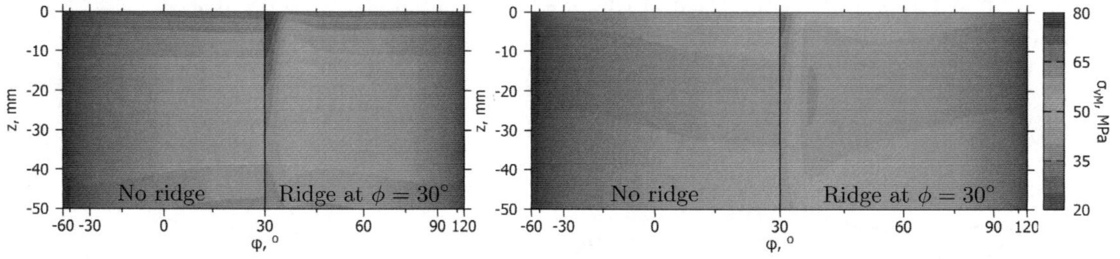

(b) ⟨1 1 1⟩, $v = 2.5$ mm/min

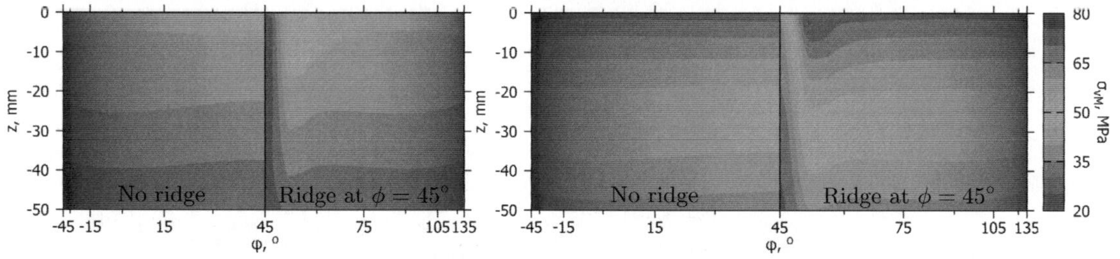

(c) ⟨1 0 0⟩, $v = 3.5$ mm/min

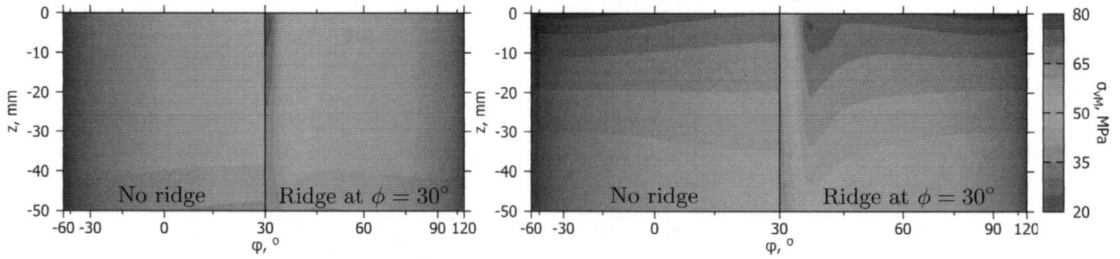

(d) ⟨1 1 1⟩, $v = 3.5$ mm/min

Figure 4. Distribution of von Mises stress on the crystal surface for different orientations and pull rates with crystal diameters of 4" (left) and 5" (right). Comparison between crystals without and with ridge. $z = 0$ corresponds to the triple point. Unified colour scale for all of the cases with 5 MPa interval between isosurfaces.

Literature data suggest that cleavage in silicon crystals occurs along the {1 1 1} slip planes [19]. These planes for ⟨1 0 0⟩ and ⟨1 1 1⟩ crystal orientations are illustrated in Fig. 6 where the angle between the planes and the vertical is 54.7° for ⟨1 0 0⟩ and 70.5° as well as 0° for ⟨1 1 1⟩

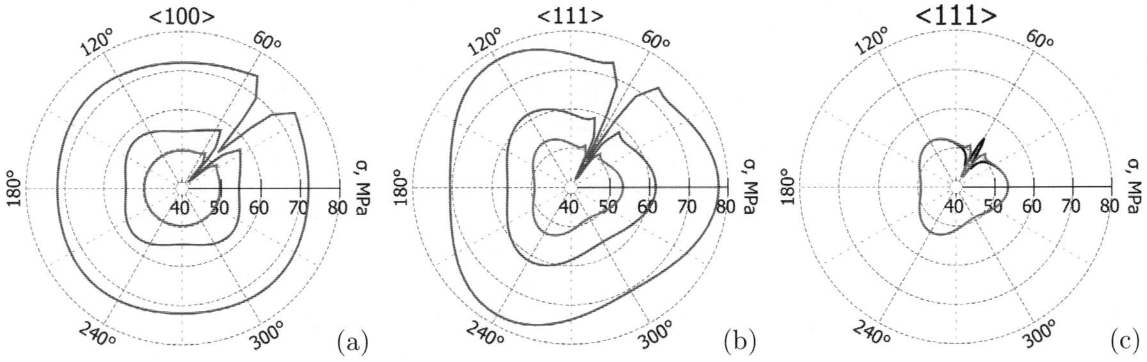

Figure 5. Azimuthal distribution of von Mises stress on the triple point line for ⟨1 0 0⟩ (a) and ⟨1 1 1⟩ (b, c) orientations. Cases of 5" crystal diameter with a straight ridge, pull rate $v = 2.5\,\mathrm{mm/min}$ without a reflector (green lines), with a reflector (blue lines) and increased pull rate $v = 3.5\,\mathrm{mm/min}$ without a reflector (red lines). Comparison between the straight ridge and undulated ridge (black line) (c).

orientation. For a quantitative measure, the normal stress on the slip planes, $\sigma_\mathrm{n} = \sigma_{ij} n_i n_j$, is examined. Distribution of the normal stress on 5" crystal with pull rate $v = 2.5\,\mathrm{mm/min}$ in Fig. 7 presents all of the possible cases for the two different crystal orientations depending on the slip plane direction. Higher normal stress for ⟨1 1 1⟩ compared to ⟨1 0 0⟩ crystal orientation is found. The stress is significantly reduced on the ridge. Strongest stress maxima are found on the crystal surface besides the ridge which exceed the values for crystal without ridge. Fracture of the crystal is the least likely to happen along the horizontal slip planes for ⟨1 1 1⟩ crystal as the normal stresses on these planes are at least an order of magnitude lower than for the other directions of the slip planes.

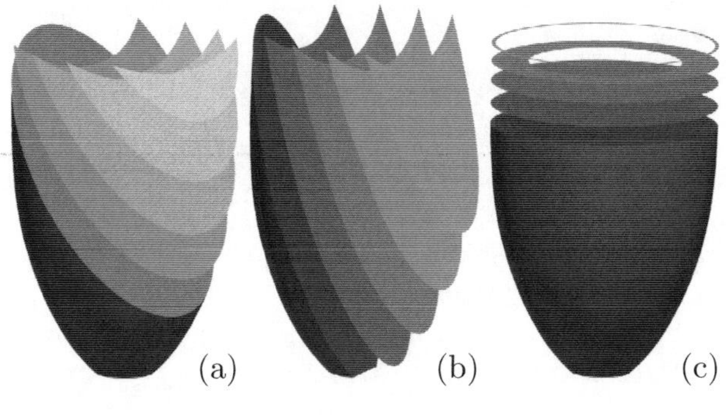

Figure 6. Illustration of the {1 1 1} slip planes. In ⟨1 0 0⟩ silicon crystal all four equivalent plane directions make 54.7° angle with vertical (a), in ⟨1 1 1⟩ crystal three equivalent plane directions make 70.5° angle with vertical (b) and one set of planes is horizontal (c).

Additionally, the shear stress $\tau = \sqrt{(\sigma_{ij} n_j)^2 - \sigma_\mathrm{n}^2}$ on the {1 1 1} slip planes can be attributed to dislocation generation. Azimuthal plots of the normal and shear stresses on the slip planes along the triple point line are given in Fig. 8 with the same order of slip plane directions as in Fig. 7. These stress projections exhibit more non-uniform behaviour compared to von Mises stress. Shear stresses are smaller than the normal stresses except for the horizontal slip planes with negligible stress. Considerable difference in the magnitude of the stress is between 4" and 5" diameter crystals. The vertically undulated ridge in the ⟨1 1 1⟩ case increases the normal stress at its peak.

Intrinsic point defects – vacancies and self-interstitials – are formed during the crystal growth which in turn lead to grown-in defects affecting the crystal quality. A criterion of (v/G), where v is the growth rate and G is the axial temperature gradient, determines the majority species

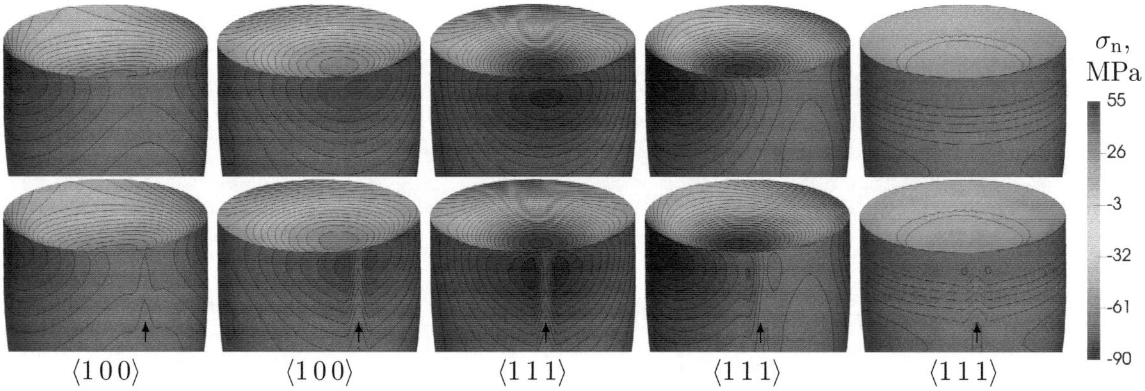

Figure 7. Distribution on the normal stress on $\{1\,1\,1\}$ slip planes for a 5" diameter crystal with pull rate $v = 2.5\,\text{mm/min}$ without ridge (top) and with ridge (bottom) for two different growth orientations. Arrow denotes position of the ridge. Plane normals from left to right: $\left(\frac{1}{\sqrt{3}}, \frac{1}{\sqrt{3}}, \frac{1}{\sqrt{3}}\right)$ or $\left(\frac{1}{\sqrt{3}}, -\frac{1}{\sqrt{3}}, \frac{1}{\sqrt{3}}\right)$, $\left(-\frac{1}{\sqrt{3}}, \frac{1}{\sqrt{3}}, \frac{1}{\sqrt{3}}\right)$ or $\left(-\frac{1}{\sqrt{3}}, -\frac{1}{\sqrt{3}}, \frac{1}{\sqrt{3}}\right)$, $\left(-\frac{\sqrt{2}}{3}, \sqrt{\frac{2}{3}}, \frac{1}{3}\right)$, $\left(-\frac{2\sqrt{2}}{3}, 0, -\frac{1}{3}\right)$, $(0, 0, 1)$.

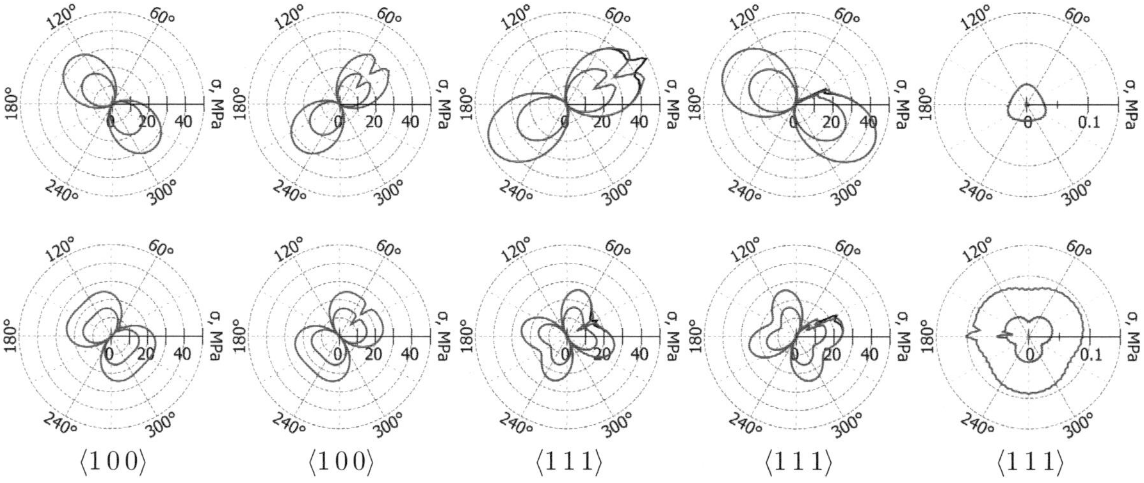

Figure 8. Azimuthal distribution of the normal stress (top) and the shear stress (bottom) on the $\{1\,1\,1\}$ slip planes along the triple point line for 4" crystal (blue lines) and 5" crystal (red lines for straight ridge and black lines for ridge with undulations) with pull rate $v = 2.5\,\text{mm/min}$. $\langle 1\,0\,0 \rangle$ orientation with the ridge at 45° and $\langle 1\,1\,1 \rangle$ orientation with the ridge at 60°. Plane normals in the same order as in Fig. 7.

according to [20]. A critical value of (v/G) separates the interstitial-rich and vacancy-rich growth modes. Influence of the thermal stress on the $(v/G)_c$ has been shown [2]. Kulkarni parameter set [21] is used to calculate the critical value:

$$(v/G)_c = 1.5582 \cdot 10^{-1} - 1.48 \cdot 10^{-3} p + 10^{-5} p^2 \tag{9}$$

where units for $(v/G)_c$ are $\text{mm}^2/\text{min/K}$ and p is the pressure in MPa. See Fig. 9 for radial distribution on the crystallization interface of ratio of the critical value with and without the stress effect. When thermal stress is included in the calculation of the $(v/G)_c$, the crystal tends to be more vacancy rich from the centre until 80% of the radius value and more self-interstitial rich in the outer part. The results for each case slightly differ between the two crystal orientations at the centre and become almost equal at the triple point. The relative changes of the $(v/G)_c$ are up to 30% of the value without stress effect taken into account in both directions of the constant value. The local effect of ridge is observed only in the last 2 mm before the triple point.

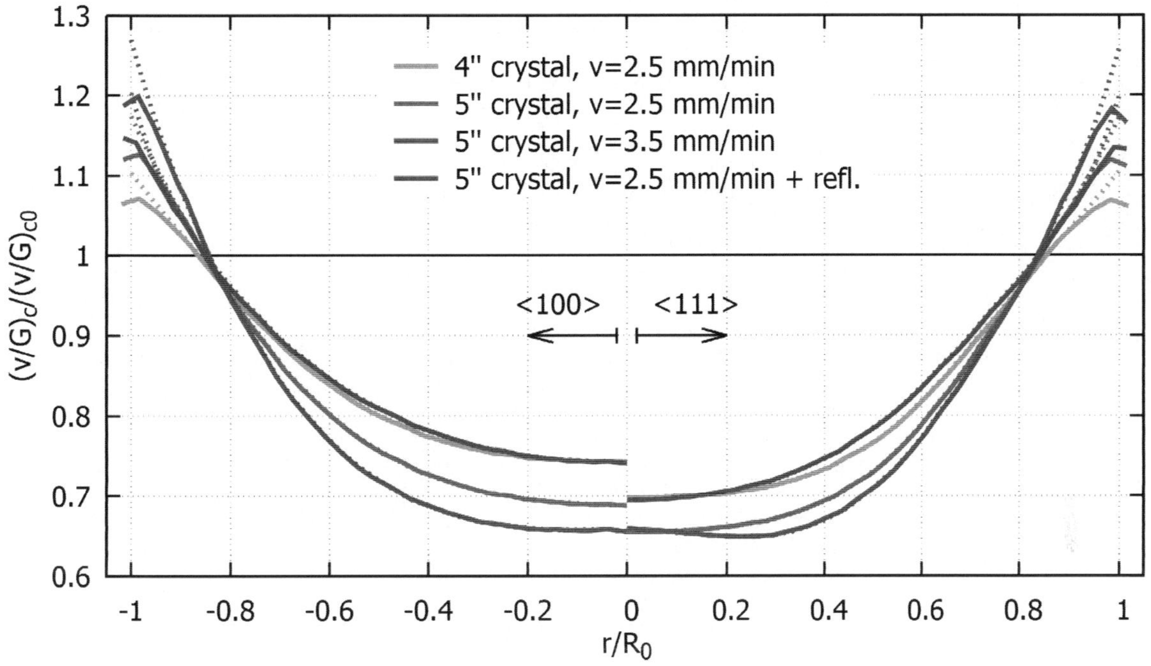

Figure 9. Normalized radial distribution of ratio between the critical values of v/G with the actual stress influence and with zero stress on the crystallization interface for $\langle 1\,0\,0 \rangle$ orientation on the left and $\langle 1\,1\,1 \rangle$ on the right. The dashed and solid lines correspond to cases without and with the ridge respectively.

4. Conclusions

Considering that ridges on the crystal surface are an essential part of dislocation-free growth, the present study shows the possible influence of the ridge on thermal stress for different process parameters. A ridge on crystal surface has been resolved by refining the mesh locally and calculations using the 3D anisotropic elasticity model reveal strong influence within the perturbed region. The stresses always decrease at the top of a straight ridge and have symmetric maxima on the sides but can also increase at the centre of the ridge with undulations. The stress projections on the slip planes have been obtained and show larger normal stresses compared to the shear stresses by approximately 8 MPa for $\langle 1\,0\,0 \rangle$ and 20 MPa for $\langle 1\,1\,1 \rangle$ orientation. Sides by the ridge are critical spots for initiation of dislocation generation and fracture as the stresses there can exceed the values for a perfectly round crystal by roughly 5 MPa (up to 8%). With the chosen reflector configuration, the thermal stress is larger compared to the standard case by 6 MPa on average and also azimuthal variations increase. Although the ridge has a local influence, the azimuthal span of the affected region on the crystal surface near the triple point line is approximately 30°. The ridge is shown to decrease the critical v/G on the crystallization interface within few millimetres from the triple point but the change is smaller compared to the overall changes due to the stress effect which range from 0.7 at the centre to 1.3 at ETP times the level without stress.

IOP Conf. Series: Materials Science and Engineering **355** (2018) 012005 doi:10.1088/1757-899X/355/1/012005

References

[1] Menzel R 2013 Ph.D. thesis The Technical University of Berlin

[2] Sueoka K, Kamiyama E and Vanhellemont J 2013 *J. Cryst. Growth* **363** 97–104

[3] Muižnieks A, Virbulis J, Lüdge A, Riemann H and Werner N 2015 *Handbook of Crystal Growth (2nd Edition)* vol 2A ed Rudolph P (Boston: Elsevier) pp 241–79

[4] Riemann H, Lüdge A, Böttcher K, Rost H J, Hallmann B, Schröder W, Hensel W and Schleusener B 1995 *J. Electrochem. Soc.* **142** 1007–14

[5] Miyazaki N, Uchida H, Munakata T, Fujioka K and Sugino Y 1992 *J. Cryst. Growth* **125** 102–11

[6] Muižnieks A, Raming G, Mühlbauer A, Virbulis J, Hanna B and Ammon W 2001 *J. Cryst. Growth* **230** 303–13

[7] Wünscher M, Menzel R, Riemann H and Lüdge A 2014 *J. Cryst. Growth* **385** 100–5

[8] Noghabi O A, M'Hamdi M and Joma M 2013 *J. Cryst. Growth* **362** 312–8

[9] Noghabi O A and MHamdi M 2014 *Journal of Crystal Growth* **400** 1–6

[10] Ciszek T 1971 *J. Cryst. Growth* **10** 263–8

[11] Fritzler K B, Trukhanov E M, Kalinin V V, Smirnov P L, Kolesnikov A V and Vasilenko A P 2007 *Tech. Phys. Lett* **33** 521–3

[12] Barinovs G, Sabanskis A and Muižnieks A 2014 *J. Cryst. Growth* **401** 137–40

[13] Lüdge A, Riemann H, Wünscher M, Behr G, Löser W, Muiznieks A and Cröll A 2010 *Crystal Growth Processes Based on Capillarity* ed Duffar T (Wiley) pp 203–275

[14] Driķis I, Plāte M, Seņņikovs J and Virbulis J 2017 *J. Cryst. Growth* **474** 8–15

[15] Fritzler K B, Trukhanov E M and Kalinin V V 2015 *Tech. Phys. Lett* **41** 731–3

[16] Ratnieks G, Muižnieks A and Mühlbauer A 2003 *J. Cryst. Growth* **255** 227–40

[17] Muižnieks A, Lācis K, Rudevičs A, Lācis U, Sabanskis A and Plāte M 2010 *Magnetohydrodynamics* **46** 475–86

[18] Swarnakar A K, Van der Biest O and Vanhellemont J 2014 *Phys. Status Solidi C* vol 11 ed Cristiano F, Pichler P, Tavernier C and Windl W (Wiley-VCH) pp 150–5

[19] Cramer T, Wanner A and Gumbsch P 2000 *Phys. Rev. Lett.* **85** 788–91

[20] Voronkov V 1982 *Journal of Crystal Growth* **59** 625–643

[21] Kulkarni M 2006 *Semiconductor Silicon 2006, ECS Transactions* vol 11 ed Huffetal H R (Pennington) p 213

IOP Conf. Series: Materials Science and Engineering **355** (2018) 012006 doi:10.1088/1757-899X/355/1/012006

Aspects of rf-heating and gas-phase doping of large scale silicon crystals grown by the Float Zone technique

F Zobel[1], F Mosel[2], J Sørensen[2], P Dold[1]

[1]Fraunhofer CSP, Otto-Eißfeldt-Str. 12, 06120 Halle, Germany;
[2]PVA Crystal Growing Systems GmbH, Im Westpark 10-12, 35435 Wettenberg,

frank.zobel@csp.fraunhofer.de

Abstract. Float Zone growth of silicon crystals is known as the method for providing excellent material properties. Basic principle of this technique is the radiofrequency induction heating, main aspects of this method will be discussed in this article. In contrast to other methods, one of the advantages of the Float Zone technique is the possibility for in-situ doping via gas phase. Experimental results on this topic will be shown and discussed.

1. Introduction

For silicon mono-crystals, the Float Zone (FZ) technique provides material of maximum quality [1–3]. Compared to the Czochralski (Cz) method, the only alternative for large diameter (<4 inch) silicon single crystal growth, the Float Zone material is 2 to 3 orders of magnitude lower in oxygen [4] and shows lower levels of metallic impurities since it is a truly crucible-free technique. The silicon is molten by a radio-frequency inductor placed near the feed rod [5, 6]. FZ silicon is the material of choice for power electronics like thyristors, insulated-gate bipolar transistors (IGBT) or any application where oxygen would lower the performance. Today, the standard diameter is 4" to 6", but 2" or 3" are available as well on the market. Due to physical limitations the maximum diameter is of about 8". Only a few companies world-wide are able to process near this limit.

In the following, important aspects on the basic principles of the FZ technique, i.e. the radiofrequency induction heating, will be discussed. Using the FZ method offers the possibility for in-situ doping via gas-phase. In contrast to other growth methods, uniform axial doping profiles can be produced, by in-situ doping via gas phase during crystallizations. Experimental results together with theoretical considerations will be shown.

2. Experimental setup

The experimental arrangement is shown in Fig.1. The setup is rather simple but the requirements for the mechanical stiffness, the quality and stability of the rf-generator for inductive heating and the accuracy and smoothness of the pulling spindles are very high since the process is very sensitive to vibrations or any sudden impacts. In addition, the specification for the feed rod material is significantly more vigorous than for any other growth techniques [4].

For the experiments, two different Float Zone machines have been used; a smaller one (FZ-14, PVA TePla) for ingots up to 4" and a large one (FZ-35, PVA TePla) for ingots up to 6". For diameters larger than 4" the argon inert gas pressure must be increased, our machine can handle an overpressure of 5 bar. According to the Paschen law, the overpressure reduces the risk of argon ionization caused by the high frequency voltage. Argon ionization is a serious problem and is one of the limitations for larger diameters.

Content from this work may be used under the terms of the Creative Commons Attribution 3.0 licence. Any further distribution of this work must maintain attribution to the author(s) and the title of the work, journal citation and DOI.
Published under licence by IOP Publishing Ltd

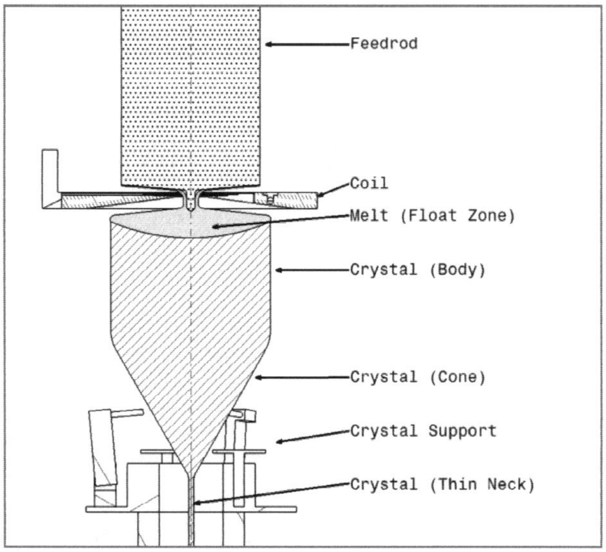

Figure 1. Cross sectional view of the Float-Zone Process. The mono crystal and the feed rod are rotating in opposite directions. The silicon melt flows as a thin film along the bottom side of the feed rod and forms a liquid bridge with the melt lake on top of the growing ingot.

2.1. Radiofrequency Induction Heating

The inductive heating of the silicon feedrod is done with a flat ("pancake") coil with one convolution. It is also referred as "needle eye" inductor, since the inner hole diameter is much smaller than the source rod and the crystal. The radiofrequency current flows with a frequency of 2,5 to 3 MHz and heats the surface of the silicon. Heating at lower frequencies (< 2 MHz) had been analysed, but shows several problems at melting the source rods, mostly the formation of solid silicon spikes [5].

Due to the skin-effect only a small layer of 270 – 290 µm on the top of the molten silicon is receiving the impact of the dynamic magnetic field [6]. Eddy currents are generated in this layer. The high thermal conductivity of silicon ($\lambda = 148 \frac{W}{m\,K}$) allows the local heat to spread very fast into the bulk material. The skin depth δ_{Sk} can be calculated by

$$\delta_{sk} = \sqrt{\frac{\rho}{\pi * \nu * \mu_0 * \mu_{rel}}} \tag{1}$$

Whereas ρ is the specific resistivity, ν the frequency, μ_{rel} the relative permittivity (for silicon melt ≈ 1,2) and μ_0 the absolute permittivity. Since the conductivity of silicon depends on the amount of base doping and temperature, a detailed look at different cases is necessary.

At 20 °C, undoped solar grade silicon shows a specific resistivity of ≥ 100 Ωcm, electronic grade silicon which comes at even higher purity can reach up to 30 kΩcm of specific resistivity. Heating the silicon leads to a rise in conductivity which also has a strong impact on the penetration of the electrical field. As shown in Fig. 2 the temperature has to rise over a certain level, for the intrinsic carrier mobility to be high enough to enable the required conductivity for inductive heating. The critical temperature is in the range of 400 to 500 °C depending on base doping and size of the specimen [7].

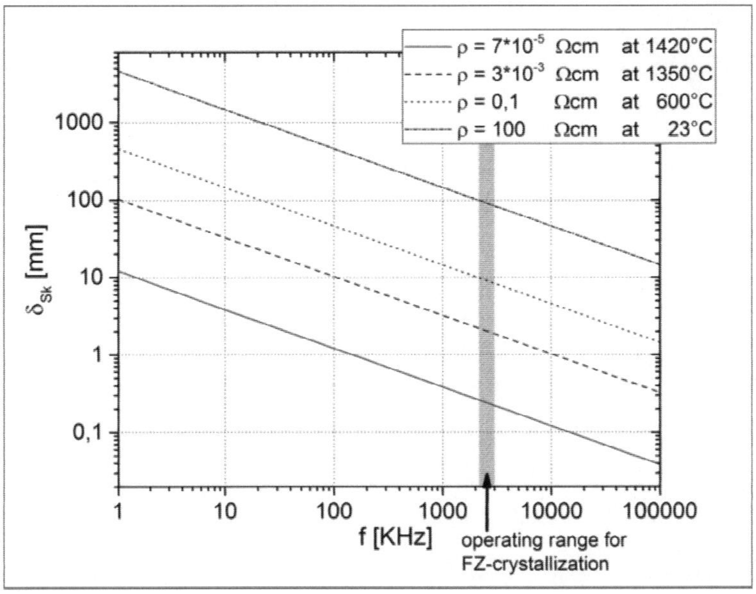

Figure 2. Skin depth depending on used frequency for rf-heating of undoped solar silicon with (R = 100 Ωcm at room temperature). The Resistivity of silicon drops with raising temperature. Resistivity calculated after [7].

To reach this temperature a graphite ring is used to preheat the silicon above this threshold value. Once molten, the conductivity of silicon remains mainly constant, a small change of about 1% can occur [8] but is negligible for further examinations. The efficiency of heating is influenced by the electromagnetic coupling; this can be improved by the shape of the induction coil that follows the melting front as close as possible.

2.2. Doping via gas phase

To influence the resistivity during crystallization an addition of doping gas is required. Phosphine for n-type and diborane for p-type are the common doping gases. For efficient incorporation the doping gas molecules must be in contact with the melt so they are blown directly on the liquid silicon surface. To ensure enrichment in the melt, the doping gas flow must overcome the natural convection in the process chamber as it is shown in Fig. 3. Due to the hot surface of crystal, source rod and the floating zone the surrounding argon gas is heated up as well and streams upwards with maximum velocities of more than 1 m/sec in the inner hole of the inductor where the gas flow is of a turbulent nature. A more laminar and steady upwards flow can be expected on the outer side of the cylindrical crystal and source rod. Streaming velocity is lower in this area. Below the surface of the outer part of the coil, the gas flow is directed radially outwards. Since the optimum contact distance and time of doping gas with the melt surface should be as long as possible, the inlet of doping gas must be pointed crosswise against the natural direction of the stream, so it can reach the free melt surface and stream along it to the inner side of the coil and then along the melting interface over the coil.

To maintain this goal, a certain pressure is required. It translates to a high inlet stream velocity. Therefore the doping gas gets diluted with argon and is blown with an overpressure of $\geq$ 0,5 bar into the chamber. The flow rate is 200 to 300 ml/min of a mixture consisting of 2 l/min argon gas and 10 to 50 ml/min of diluted doping gas. Since the doping gas is diluted in argon at 100 ppm the finally used gas concentration is $\approx$ 2,5 ppm. All flow rates are given in norm (milli)litres per minute.

The setup used at the Fraunhofer CSP consists of a small tube with 3 mm inner diameter. The distance between outlet and crystal is about 30 mm. The gas outlet is located at the level of the crystallization interface and points slightly upwards (Fig. 3).

Doping efficiency η_D can be calculated after:

$$\eta_D = \frac{a\dot{t}_C}{n * a\dot{t}_G} = \frac{\dot{V}_C * C_C}{\dot{V}_G * C_G * \frac{1}{V_m} * N_A} = \frac{\emptyset_C^2 * \frac{\Pi}{4} * v_c * C_C}{\dot{V}_G * C_G * \frac{1}{V_m} * N_A} \quad (2)$$

Where the ratio $\frac{a\dot{t}_C}{n * a\dot{t}_G}$ stands for the flow of atoms that are incorporated into the crystal divided by the flow of dopant atoms going into the chamber. The factor $n = 2$ for diborane (B_2H_6) and n=1 for phosphine (PH_3), since diborane consists of two boron atoms per molecule. $\dot{V}_C$ is the volume flow of crystalline silicon that is calculated based on the diameter of the crystal $\emptyset_C$ and the crystallization rate v_c. It is multiplied with the concentration of dopant atoms in the crystal C_C, that can be calculated from the crystal resistivity [9]. $\dot{V}_G$ stands for the gas volume flow of the doping gas, C_G for the concentration of pure doping gas in the mixture (in this case ≈100 ppm), V_m is the molecular volume and N_A the Avogadro constant.

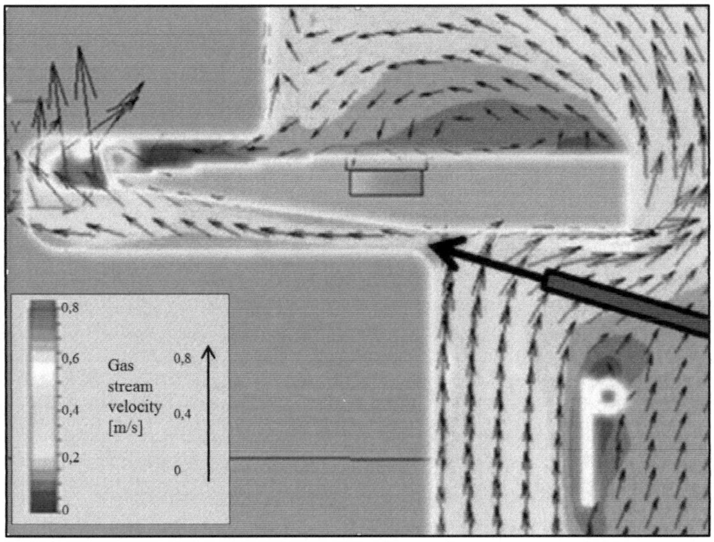

Figure 3. Positioning of doping gas inlet; at a high flow velocity the doping gas stream perpendicular to the natural convection to the free melt surface then it is taken along with the gas flow along the silicon surface through the coil hole in an inwards direction, and then outwards along the melting front for efficient incorporation. (CFD simulation carried out with FlowEFD by PVA Tepla).

3. Results

3.1. P-type doping with diborane (B_2H_6)

Investigations of doping of a 5 inch crystal via gas phase and diborane were conducted. The diborane gas flow was enabled after the starting cone of the crystal and set to a value of 50 ml/l, with an argon flow rate of 2 l/min and an overpressure of 0,5 bar (Fig.4). The specific resistivity of the grown crystal was measured with an eddy current sensor and cross checked via four point probe measuring. The crystal position could be determined by assigning it to the lower spindle position that is logged during the growth process.

The specific resistivity at the beginning of the crystal (up to 300 mm length, see Fig. 4) was about 600 Ωcm, an undoped source rod was used for crystallization. Immediately after enabling the gas flow a drop in resistivity of the crystal to 50 Ωcm is observed. That equals a rise in boron concentration of the crystal form 2,2 * 10^{13} at/cm³ to 2,6 * 10^{14} at/cm³. As can be seen in Fig. 4 an asymptotic drop to 7 Ωcm (equals 1,9 * 10^{15} at/cm³) follows. It takes 100 mm of grown crystal length to reach this equilibrium state.

The ratio of boron atoms blown into the chamber and resulting boron atom concentration in the crystal equals to 24,5 %, considering the crystallization velocity of 3 mm/ min. Nearly every 4th atom of boron that is blown into the chamber is incorporated into the melt. The rather low value is not surprising, since diborane is thermal unstable an starts decomposing already at 100 °C in hydrogen and other borane complexes like B_4H_{10}, B_5H_9, B_5H_{11}, $B_{10}H_{14}$ or higher molecular solid yellow boron $(BH_{\approx 1})_x$ [10]. After a longer process time elementary boron can be seen as dark brown film covering the gas nozzle.

3.2. N-type doping with phosphine (PH3)

Just as the p-type doping, the n-type variation was analyzed at a 5 inch crystal which was grown with a speed of 3 mm/min. The starting cone was grown without doping to have a reference value of specific resistivity. The source rod was different to the first one but undoped as well. It resulted in a specific crystal resistivity of 120 Ωcm (3,6 * 10^{13} at/cm³). In this experiment the amount of phosphine gas flow rate was ramped up in three steps, from 0 to 30, 40 and finally 50 ml/min. This can be seen in Fig 5. The corresponding resistivity measured at 12, 7, and 3 Ωcm (3,7 * 10^{14} at/cm³; 6,4 * 10^{14} at/cm³ and 1,5 * 10^{15} at/cm³). The absorption rate was determined based on this values is 42,5 %, nearly every second phosphorus atom is incorporated in the crystal. Phosphine is more thermally stable than diborane, its decomposition temperature is about 500 °C [11]. Better doping efficiency was expected and could be proven. Since the necessary amount of dopant atoms in silicon is lower in n-type due to the higher mobility of the free electrons in the crystal lattice, a lower amount of phosphorus is required for the same resistivity compared to p-type.

Still, the amount of incorporated atoms is low, compared to values from modelling of a 4 inch process that predict a dopant efficiency of up to 95 % [12]. Reasons for these differences might be the evaporation of phosphorus due to the high partial pressure, but this is supposed to be rather low due to the high chemical affinity of phosphorus and silicon. Another difference that might explain the deviations is the coil designs. A relatively small inductor hole compared to the simulated process in [12] might supress the gas flow next to the floating zone. The dopant concentration in the gas flowing along the melting surface in the area above the coil might be lowered, due to this reason.

For both, p- and n-type doping the incorporation ration of dopants might be increased with a higher outlet velocity and less distance. The inlet flow rate can be increased by using a higher doping gas pressure. A smaller diameter of the gas inlet or a nozzle with reduced exhaust area even increases the doping gas inlet velocity and can improve the amount of doping gas molecules reaching the melt surface. The distance can be reduced as well, but to prevent arcing, the tube or nozzle should be of an insulating material, e.g. quartz, that is also resistant to the occurring high temperatures from the radiating silicon melt and crystal.

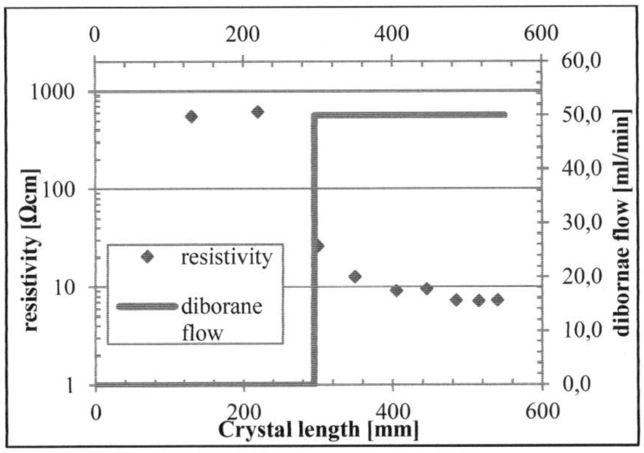

Figure 4. P-type doping via gas phase with diborane resistivity of a 5 inch FZ-crystal (grown with 3,0 mm/min) in axial direction.

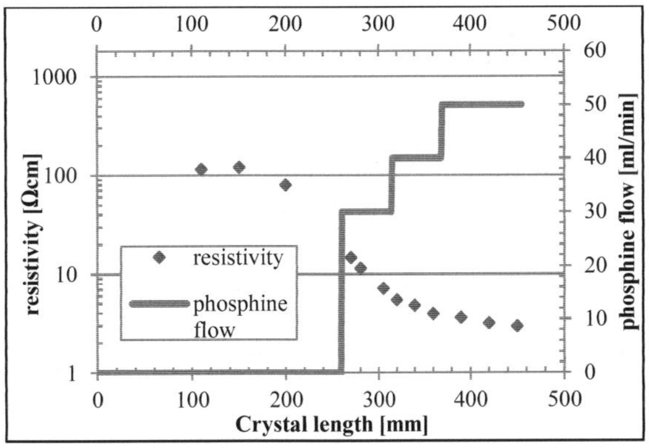

Figure 5. N-Type doping via gas phase with phosphine resistivity of a 5 inch FZ-Crystal (grown with 3,0 mm/min) in axial direction.

4. Conclusions

In the article we have shown important aspects and considerations on radiofrequency induction heating. The temperature dependence of the conductivity of silicon, and its impact on the skin depth was discussed. This leads to the requirement of high working frequencies.

Experiments on gas-phase doping have been carried out, showing a doping efficiency of 25 % (43 %) for diborane (phosphine) in the current experimental setup. The influence of the gas-flow on the resistivity has been shown. The reached efficiency shows a much smaller value than simulations suggest, deviations are discussed. A smaller tube or a nozzle in closer proximity to the melt with higher doping gas pressure might increase the ration of doping atoms blown in the chamber to doping atoms incorporated in the crystal. Further investigations will provide additional data to confirm these assumptions.

Acknowledgement

The authors gratefully acknowledge the financial support provided through the BMWi (Bundesministerium für Wirtschaft und Technologie) within the project no. 0325822D KosmoS.

References

[1] Rost H J, Luedge A, Riemann H, Kirscht F and Schulze F W 2012 *Crystal Research and Technology* **47** 273–8
[2] Ciszek T F and Wang T H 2000 *Proceedings of SPIE - The International Society for Optical Engineering* **4218** 105–17
[3] Dold P 2015 *Semiconductors and Semimetals* **92** 1–61
[4] Zobel F, Dold P, Kunert R, Lauer K and Michl B 2014 *Float Zone Kristallzüchtung für Solare Anwendungen (Deutsche Kristallzuechtungstagung)* (Halle (Saale))
[5] Rost H J, Menzel R, Luedge A and Riemann H 2012 *Journal of Crystal Growth* **360** 43–6
[6] Hurle D T J (ed) 2016 *Bulk Crystal Growth: A. Basic Techniques (Handbook of Crystal Growth)* (Saint Louis: Elsevier Science)
[7] Li S S 1977 *The dopant density and temperature dependence of electron mobility and resistivity in n-type silicon (Semiconductor Measurement Technology* 400-33*)* (Washington, D.C.: National Bureau of Standards)
[8] Sasaki H, Ikari A, Terashima K and Kimura S 1995 *Japanese Journal of Applied Physics, Part 1: Regular Papers and Short Notes and Review Papers* **34** 3426–31
[9] National Bureau of Standards 1981 The relationship between resistivity and dopant density for phosphorus-and boron-doped silicon *(*400-64*)* (Washington, D.C.: U.S. Department of Commerce) (accessed 13 Jan 2017)
[10] Holleman A F, Wiberg E and Wiberg N 2007 *Lehrbuch der anorganischen Chemie* 102nd edn (Berlin: De Gruyter)
[11] Baillargeon J N, Cheng K Y, Jackson S L and Stillman G E 1991 *Journal of Applied Physics* **69** 8025–30
[12] Sabanskis A, Surovovs K and Virbulis J 2017 *Journal of Crystal Growth* **457** 65–71

IOP Conf. Series: Materials Science and Engineering **355** (2018) 012007 doi:10.1088/1757-899X/355/1/012007

Development of silicon growth techniques from melt with surface heating

Anatoly Kravtsov

KEPP EU AS, Carnikavas iela, 5, Riga, LV 1034, Latvia

E-mail doc@keppeu.lv

Abstract. The paper contains literary and personal data on the development history of silicon-growing technology with volumetric and surface melt heating. It discusses the advantages and disadvantages of surface-heating technology. Examples are given of the implementation of such processes in the 60s-70s of the last century, and the reasons for the discontinuation of the relevant work. It describes the main solutions for the implementation of crystal growth process with the electron-beam heating of the melt surface, implemented by KEPP EU (Latvia). It discusses differences in the management of the growth process for the crystals with constant diameters compared to the Czochralski method. It lists geometrical and electro-physical properties of the obtained crystals. It describes the possible use of such crystals and the immediate challenges of technology development.

1. Introduction

The historical overview of silicon crystal growth methods goes beyond the scope of this article, but in my opinion, since some historical data corroborate the ideas presented, they should be illustrated in Table 1.

Table 1. The historical overview of silicon crystal growth methods.

Tab 1. Silicon crystal growth history

Year	1950	1956	1962	1967	1972	1973	1974	1976	1980	1984	1986	1988	1990	1992
CZ diameter, mm [1]	12	25	40	50	65	78			100	125		150		200
CZ charge, kg [1]	0,05	0,4	1,2	2,5	6	12			24	38		65		110
FZ diameter, mm [2, own data]		10	40				65	78	100		125		150	
FZ/pydestal diameter [3,4]			40		42									
Skull silicon diameter,mm [5, own data]				40	45									

2. The influence of heating methods on the properties of crystals and the development of processes

The fundamental difference between the CZ (Czochralski method) and DS (direct solidification) methods from the FZ process, the pedestal growing, and the skull-melting is most vividly illustrated by the numerical simulation diagrams of these processes shown in Figure 1. The typical size of the heating zone in FZ processes and when growing by electron-beam heating is 10-30 mm. For growing methods with volumetric melt heating (CZ and DS), the typical size of the heating zone is measured in

Content from this work may be used under the terms of the Creative Commons Attribution 3.0 licence. Any further distribution of this work must maintain attribution to the author(s) and the title of the work, journal citation and DOI.
Published under licence by IOP Publishing Ltd

hundreds of millimetres, and the surface area in modern systems is about 1 square metre. It is clear that in the case of volumetric heating, the systems are characterised by more homogeneous thermal fields favourable to the growth of crystals. At the same time, a large number of different tooling mainly made of graphite used in volumetric heating methods results in inevitable contamination of the melt during the process. The absence of such a tooling in the processes of surface heating provides these methods not only with the ability to produce cleaner crystals, but also gives them some economic advantage due to the absence of the expenses for the consumable tooling.

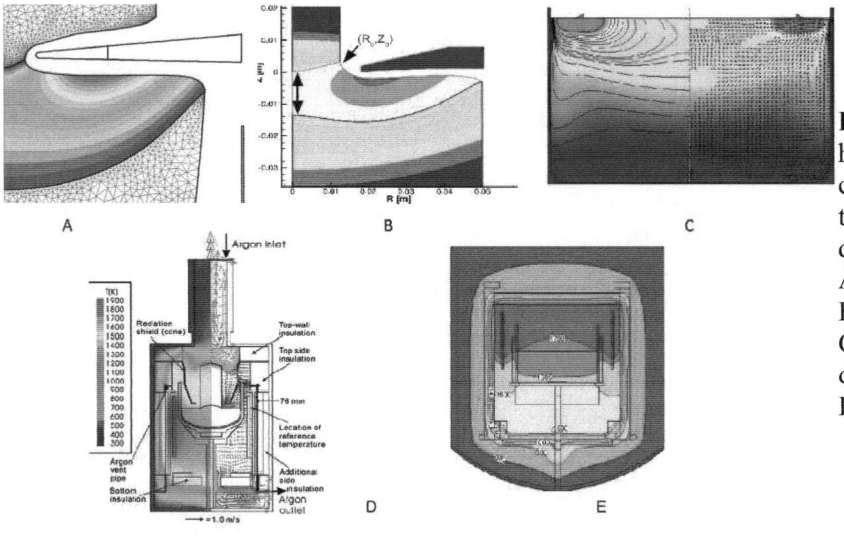

Figure 1. Area of heating in different crystal growing techniques (temperature distribution):
A- FZ [8];
B - pydestal [4];
C - Electron beam [own data];
D - CZ [9];
E – DS [10].

On the other hand, surface heating methods have a number of drawbacks.

For the FZ process and pedestal growing, a polycrystalline silicon rod is needed, corresponding to a number of specific geometry requirements - high density, and lack of cracks and amorphous phase inclusions. The fact is that the diameters of the original and growing rods should not differ by more than 20% in the standard FZ process. As for the pedestal growing process, the situation is even worse - the diameter of the growing rod is almost 2 times smaller than the pedestal diameter. The development of deposition technology for these rods in the Siemens process takes considerable time. As a result, the lack of adequate methods of obtaining rods hindered the development of the FZ process in the late 1960s, in the early 1970s, and in the early 1990s. To ensure the quality requirements for the rods, the processes of their deposition are executed at a slow rate. Compliance with this growing technology makes rods very expensive - at least twice as much as pieces of polycrystalline silicon of the relevant quality used in the process of growing from the crucible. The high price of rods causes a high price of FZ single crystals and limits their use. This leads to another problem - the complex and costly equipment for the FZ process is produced in small amounts.

The high cost of developing such equipment is allocated to a small number of produced plants and significantly increases the cost of equipment. As a result, a second significant component of the cost of FZ silicon - depreciation deductions - are added, which adds to the price and restricts the use of FZ single crystals. Savings resulting from the lack of additional consumables are insufficient to compensate for the high cost of raw materials and depreciation deductions, so it results in the use of FZ silicon being limited to special detectors and power electronics, where such silicon cannot be replaced with the cheaper one. The ever-increasing demands to energy network loads add to the requirements for plate diameter.

Given the current trend and according to the estimates [12], the need for plates of 300 mm in diameter should be expected by 2020. In order to solve this problem, firstly, it is extremely important to produce the high-purity rods of the appropriate diameter (250-350 mm), and secondly, it is necessary to reduce the voltage in the inductor to eliminate the discharges typical of the more than 2

IOP Conf. Series: Materials Science and Engineering **355** (2018) 012007 doi:10.1088/1757-899X/355/1/012007

MHz current frequency used in the FZ process. When using low frequencies, such as 1.76 MHz [11], the problems occur with the source rod melting. Thus, currently there is an interest in 300-millimetre plates for powerful thyristors, but there are no polycrystalline rods for the production of single crystals with a diameter of 300 mm and equipment for the implementation of the FZ process. However, real industrial development will require a solution in the next 5-7 years. Our goals was to create equipment and technology for the production of feeding silicon rods for FZ process of growing silicon single crystal 300 mm in diameter.

3. Growing crystals with electron beam heating

Electron-beam heating processes were used in the growth of single crystals with dislocations on a pedestal [4] in a vacuum. There are two main problems with this process: the optimal distribution of heat, and the substantial evaporation of silicon. Evaporated silicon condenses on the surface of a vacuum chamber, and then the condensate falls into the melt and enters the crystallisation surface, resulting in the stop of the crystal's monocrystalline growth. As a result, these processes have not progressed. On the other hand, several versions of the process using skull-melting were patented at different times, and their diagrams are shown in Figure 2.

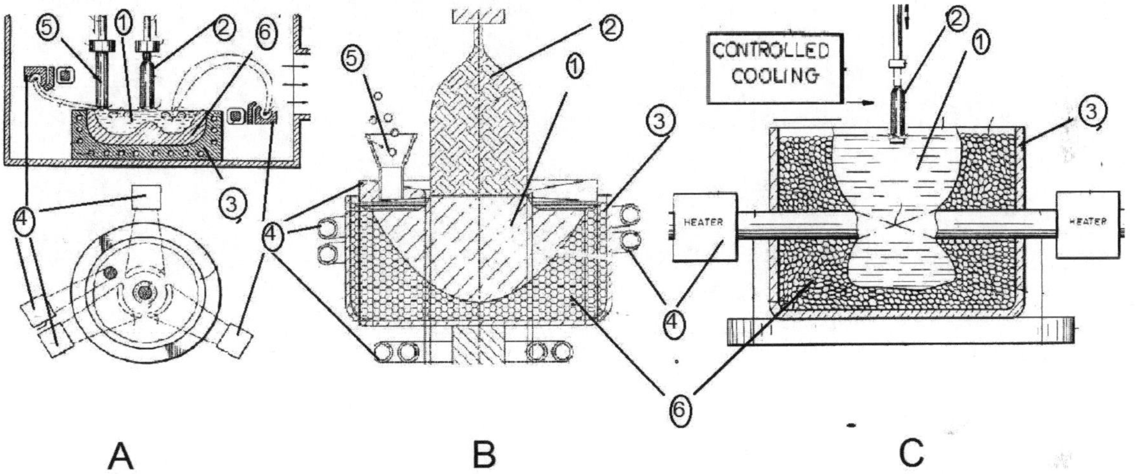

Figure 2. Different types of skull pulling processes
A–from cold crucible [5]; C–multiple induction heating [14]; B– inside electrical current heating [13]. 1- melt; 2-pulled crystal; 3-crucible; 4-heating source (A- electron beam; B- multiple induction coils; C- electrical current flow); 5- silicon source; 6 – silicon skull.

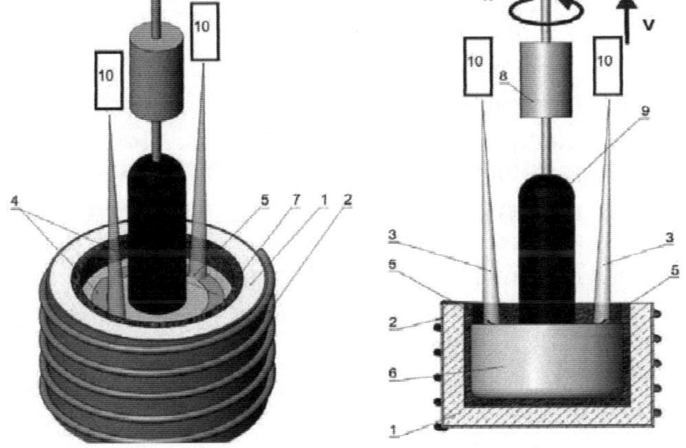

Figure 3. Electron beam pulling process developing at KEPP EU
1-cold crucible;
2- insulation;
3-electron beam;
4- way of focal spot on melt surface;
5-focal spots; 6- melt;
7- container; 8- seed holder
9 – pulling crystal;
10- electron beam gun.

IOP Conf. Series: Materials Science and Engineering **355** (2018) 012007 doi:10.1088/1757-899X/355/1/012007

Although these diagrams have a number of drawbacks, silicon crystals have been obtained using them. The diagram in Figure 2a greatly benefits from zero requirements for the size and shape of the source material, except for the feeding rod. Obviously, the main problem with the process is the difficulty of heating up the central area of melt, which is cooled using a copper crucible and tends to attach to the growing crystal. To resolve the above problem, we developed a furnace and a process diagram with reduced heat emission from the bottom part of the melt shown in Figure 3.

Thermal insulation improves the energy efficiency of the process. In the case of the rod growth with the diameter of 200 mm, the entire process consumes up to 25 kWh/kg, including melting of the source silicon and growing of the crystal; the crystal was grown at a rate of 0.5-2 mm / min while rotating at least 10 rpm. The first crystals with a diameter of about 100 mm were obtained in 2011. In 2014, the diameter of rods grown in a stable process reached 170 mm, in 2016 it was 220 mm and 300 mm in 2017.

Figure 4. Electron beam pulling process. A- puller (detail description in [15]); B- 6"; 9"; 12" ingots archived on puller shown at A; C- view of process.

Developed equipment does not use a movable melt bath, thus eliminating the mixing of melt resulting from the crucible's rotation. The symmetry of heating is ensured by the movement of the electron beams' focal spots. However, there is a problem with the control of the crystal's diameter. As the crystal grows, the melt level decreases and the controlled meniscus moves away from the camera. The attempt to use the laser distance meter to determine the position of the melt and introduce an adjustment in the direction of the camera axis has encountered a problem of reflected rays being scattered, as there is a wave on the melt surface. This wave is the natural property of the process. The point is that the electron beam creates pressure on the surface of the melt, and as a result, the beam displacement forms a wave. In addition, the electron beam produces a glowing trail, and the melt meniscus near the skull-melting surface has almost the same brightness as the control meniscus near the growing surface of the crystal (Figure 4).

As a result, the camera periodically "loses" the monitored object. We were actually able to solve this problem [16] by filtering the images. This resulted in a programme developed to adjust the diameter by changing the speed of pulling, but the control over the melt level is not yet resolved. As a result, an industrial plant designed for growing polycrystalline rods from silicon (Figure 4a) has been developed, and the methods for growing rods with a diameter of up to 220 mm have been mastered. Also, the possibility of growing crystals with a diameter of up to 300 mm (12 inches) has been demonstrated. The maximum mass of the obtained crystals is 90 kg at 110 kg of consumed material.

The results of monitoring impurities in the grown crystals (N 3-2) and melt residues (N 5-2) compared to FZ silicon of high purity (N-HR) are shown in Figure 5.

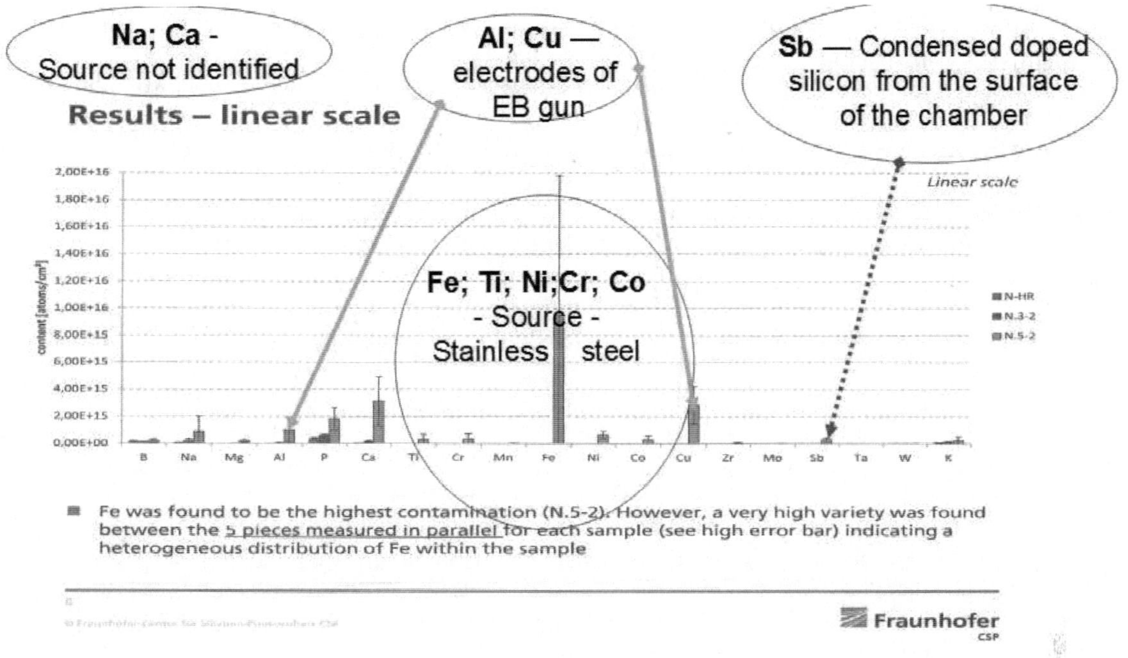

Figure 5. Results of impurities control in high resistivity FZ sample (N-HR), sample from bottom part of electron beam pulled rod (EBR) (N 3-2) and residues of melt (N 5-2).

4. Increasing the purity of silicon grown using electron beam heating

Within the capabilities of ICP MS, it has been demonstrated that the resulting purity of the rod (sample No. 3-2) is almost the same as the purity of the high-purity reference sample of FZ silicon (N-HR). The analysis of detected impurities indicates the need to clean the surface of the vacuum chamber before the process of obtaining pure rods, and to form and maintain a condensed pure layer of silicon on them. The formation of a group of metallic impurities similar to stainless steel by composition indicates that the surfaces of the vacuum chamber and other steel parts are dispersed when irradiated by the primary or secondary electrons. The most dangerous surfaces are those that can be irradiated by primary electrons with trajectories slightly deflected from the main flow axis. Because of the relatively small number of such electrons, they do not leave traces on steel surfaces but result in saturation of residual gases with metal vapours in the vacuum chamber. Another source of contamination is the electron-beam heater electrodes. In order to prevent electrode materials from entering the melt, a gas-dynamic window was developed to separate the heater's gas-discharge chamber from the plant's processing chamber. On the side of the processing chamber, an inert gas is fed to the membrane, and on the side of the gas-discharge chamber, a zone of increased pressure is created for a mixture of this inert gas with gases that initiate plasma and contain vapours of electrode materials. Along with the electrode materials, the gas mixture is removed from the high-pressure zone by an additional vacuum pump.

5. Conclusions

1) The advantages and disadvantages of silicon crystal growth technology using local surface heating are analysed.

2) The world's first equipment and technology have been developed, allowing to ensure the growth of silicon rods from the melt heated by the electron beam. An example of methods of obtaining polycrystalline rods with a diameter of up to 300 mm suitable for growing single crystals in the FZ

process is provided. The purity of rods monitored at the level of the MS ICP method corresponds to the purity of the standard FZ silicon.

3) Sources of impurities entering the melt during the process of growing have been identified, detected by analysing the traces in the melt residues left after the growing process. Ways of eliminating these sources to improve the quality of the grown rods have been designated.

Acknowledgement: The research leading to these results has been supported by the European Regional Development Fund project "Development of Competence Centre of Mechanical Engineering", contract No.1.2.1.1/16/A/003 signed between Competence Centre of Mechanical Engineering and Central Finance and Contracting Agency, Research No. 2.3 “Modeling and optimizing the form of electron beam heater plasma charge for improving the management of crystal growth processes”.

References

[1] Werner Zulehner, 3 April 2000, Historical overview of Silicon crystal pulling development. *Materials Science and Engineering*: B, Vol. 73, Issues 1–3, pp 7-15

[2] Ratnieks G., 2007, Modelling of the Floating Zone Growth of Silicon Single Crystals with Diameter up to 8 Inch. Ph.D. Thesis.University of Latvia Faculty of Physics and Mathematics Department of Physics, 176

[3] Ciszek T F, 1972, Growth of 40 mm diameter single crystals by a pedestal technique using electron beam heating, *Journal of crystal growth*, No. 12, pp 281-287

[4] Wünscher M, Riemann H, Hallmann-Seifert B, Lüdge A, March 23-30 2011, Crucible-free Crystal Pulling of Germanium Scientific Seminar on Application of Germanium Detectors in Fundamental Research Tsinghua University, Beijing/China

[5] Charles W H, Charles D A H, 1968, Method for growing crystals, Patent US 3494804 A Continuation-impart of application Ser. No. 659

[6] Schmid F, 2013, History of Technologies, development for Solar Silicon Cost Reduction, Solar Power for the World What You Wanted to Know about Photovoltaics Wolfgang Palz Pan Stanford Publishing, pp 257–274

[7] Jinggong, Science & Technology Co., Ltd.: http://www.machingservices.com/polysilicon-casting-furnace-manufacturing-products.html

[8] Muižnieks A, Rudevičs A, Krauze A, Suškovs V, Surovovs K, Janisels K, Mathematical modelling of silicon single crystal industrial growth, https://www.lu.lv/fileadmin/user_upload/lu_portal/projekti/atom/Muiznieks.pdf

[9] Chung-Wen L, Chao-Kuan H and Wen-Chin H, 2009, Czochralski Silicon Crystal Growth for Photovoltaic Applications, *Crystal Growth of Silicon for Solar Cells,* XIV, 255, pp 25-39

[10] Wei J, Zhang H, Zheng L, Wang C, Zhao B, 2009, Modeling and improvement of silicon ingot directional solidification for industrial production systems, *Solar Energy Materials & Solar Cells* 93, 1531–1539

[11] Menzel R, 2013, Growth Conditions for Large Diameter FZ Si Single Crystals von der Fakultät II – Mathematik und Naturwissenschaften der Technischen Universität Berlin zur Erlangung des akademischen Grades genehmigte Dissertation Dr. Ing. Berlin, p 101

[12] Deboy G, Aigner K, 2013, Power Semiconductors on 300-Millimeter Wafers, *Semiconductors*, Issue 3, Power Electronics Europe, pp 44-47

[13] Pinkhasov E, Vernon Mt, 1984, Method of and apparatus for the drawing of bars monocrystalline silicon, United States Patent 4,575,401 Flled: Int. c1.4 c301; 15/14 US. Cl.. 156/602; 156/617 M

[14] Riemann H, Abrosimov N V, Fischer J, Renner B M, Wusterhausen K, 2010, Method and apparatus for producing single crystals composed of semiconductor material, *US Patent Application Publication* (10) Pub. No.: US 2012/0285369 A1 Filed: Int. Cl. C30B 15/14 (2006.01) C30B 15/02 (2006.01) (52) U.S. Cl. 117/13; 117/214

[15] Kravtsov A, 2014, Ingots Pulled with Electron Beam Heating from Skull a New Feedstock for FZ Crystals Applicable for Solar Cells, in IEEE PVSC40, Denver, CO, USA
[16] Kravtsov An, Kravtsov Al, Fuksis R and Pudzs M, 2015, Ingot pulling with electron beam heating: Process enhancements, *Photovoltaic Specialist Conference (PVSC)*, IEEE 42nd, New Orleans, LA, pp 1-4

Modeling electron beam parameters and plasma interface position in an anode plasma electron gun with hydrogen atmosphere

A Krauze[1], **J Virbulis**[1] **and A Kravtsov**[2]

[1] Center for Processes Analysis and Research Ltd, Zeļļu str. 25, LV-1002 Riga, Latvia
[2] KEPP EU LLC, Carnikavas str. 5, LV-1034 Riga, Latvia

Abstract. A beam glow discharge based electron gun can be applied as heater for silicon crystal growth systems in which silicon rods are pulled from melt. Impacts of high-energy charged particles cause wear and tear of the gun and generate an additional source of silicon contamination. A steady-state model for electron beam formation has been developed to model the electron gun and optimize its design. Description of the model and first simulation results are presented. It has been shown that the model can simulate dimensions of particle impact areas on the cathode and anode, but further improvements of the model are needed to correctly simulate electron trajectory distribution in the beam and the beam current dependence on the applied gas pressure.

1. Introduction

Silicon (Si) is the most widely used semiconducting material in microelectronics, high-power electronics and solar applications.

Traditional silicon crystal growth methods have limitations that do not allow them to grow large-diameter high-purity Si monocrystals. In the Czochralski method, Si monocrystals are pulled from Si melt in a quartz crucible. Due to crucible material dissolution in the molten Si, impurities are incorporated in grown crystals. Float zone (FZ) method requires high-purity polycrystalline Si feed rods as input material, and needle-eye high-frequency electromagnetic inductors that are used for the feed rod melting make it difficult to increase the diameter of the grown crystals.

An alternative to these systems could be a silicon crystal growth system where a crystal is pulled from a crucible in which the silicon melt is heated by a high-power beam glow discharge based electron gun, see figure 1. Because the beam energy is concentrated on a very small area of the free silicon melt surface, the silicon melt can be heated locally to very high temperatures. This allows to purify the silicon material to a very high degree. On the other hand, such melt heating allows to preserve a layer of solid Si crust between the melt and the crucible, which prevents contaminants from the crucible from entering the crystal. This method therefore can potentially be used to grow both high-purity and very large diameter crystals.

A key issue of electron guns is wear and tear of gun parts (cathode and anode in particular) that is caused by impacts of high-energy charged particles produced inside the guns. Besides increased exploitation costs, such wear also serves as an additional contamination source which must be reduced. One way to reduce the wear and tear is to reduce the particle impact areas. A mathematical model that can describe the working of the electron gun and potentially show the way to optimize it would be very helpful for this purpose.

Content from this work may be used under the terms of the Creative Commons Attribution 3.0 licence. Any further distribution of this work must maintain attribution to the author(s) and the title of the work, journal citation and DOI.
Published under licence by IOP Publishing Ltd

For this reason a program system for modeling the applied beam glow discharge based electron gun has been developed. The program system simulates 2D axisymmetric EM field distribution in the gun and calculates 3D particle trajectories to model the electron beam formation. The aim of the model development is to obtain a tool that would correctly model the basic working parameters of the gun, especially the dimensions of the ion and electron impact areas on the cathode and anode. In this article, we present the description of the employed mathematical models, their numerical implementation, and the first simulation results.

2. Description of existing software for modeling electron guns

There are several software programs available for modeling electron guns. We will give here a brief description of some of the software.

EGUN/IGUN software package is a set of programs for modeling different aspects of electron guns, [1]. EGUN is a 2D electron gun simulation software that calculates relativistic electron trajectories in electric and magnetic fields. It solves Poisson's equation for the electric potential which is coupled with charge distribution generated by moving electrons. Similarly the magnetic field simulations take into account electron currents. Different modes of electron emissions are considered. IGUN is a related program for positive ion extraction from plasma, which uses a 1D model for the plasma sheath simulations. It seems however that EGUN and IGUN are two separate programs, whereas in our case we have to consider both electron and positive ion trajectories.

Opera FEA Simulation Software [2] is finite element electromagnetic simulation software that can also model electron guns. Similarly to EGUN/IGUN it also takes into account the space-charge and beam current influence on the EM field distributions and particle trajectories. This program contains additional features such as secondary particle emission from surfaces and ionization in the gas volume. A constant-potential plasma surface with constant-rate ion emission is implemented as an ion emission model for the plasma, and the plasma shape is determined iteratively. An example of a simulation of ion extraction from plasma surface can be found in [3].

Charged Particle Optics Software (CPO) is another software for calculation of electrostatic and magnetostatic fields and trajectories of charged particles flying through these fields. CPO uses a boundary element method for the field calculation, see [4].

Reference [5] describes a mathematical model of a beam glow discharge based electron gun which directly corresponds to the type of the gun that is considered in this article. As in all the previous models, it considers modeling a self-consistent electric field distribution coupled with a model for tracking charged particle movement in this field and calculation of the charge distribution. A description of boundary conditions for the electric field on the plasma boundary and condition required for finding its equilibrium shape are also given. Our mathematical model was based on this source.

3. Description of the working principles of the crystal growth facility and the applied electron gun

KEPP EU uses a beam glow discharge based electron gun for heating the free Si melt surface. The gun works by applying high voltage (30 kV) between its cathode and anode, see figure 2. A low-temperature low-ionization plasma is lit in the low-pressure (some Pa) hydrogen working gas near the anode. The positively charged hydrogen gas ions can escape from plasma into the neutral gas between the cathode and plasma. These ions, accelerated by the electric field, hit the cathode surface. As a result, cathode material atoms are released from its surface, and they can contaminate Si material.

The ion hits cause also emission of several secondary electrons. Accelerated by the electric field, they form a narrow beam that travels through a hole in the anode in the direction of the Si melt. Because the beam is larger than the hole, the anode surface is also damaged, and its material also becomes a contaminant. The hole cannot be freely widened because it will increase the gas flow out of the gun chamber and increase Si contamination.

Outside the gun chamber, the beam is affected by a system of electromagnets. Since the optimization of the beam control was not part of this task, only the two electromagnets that were the closest to the anode and could affect the electron trajectories were taken into account in the model.

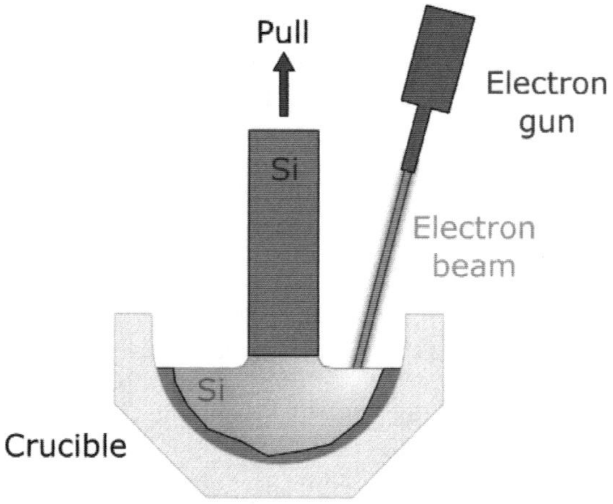

Figure 1. A general scheme of a crystal growth system that uses an electron gun for the Si melt heating.

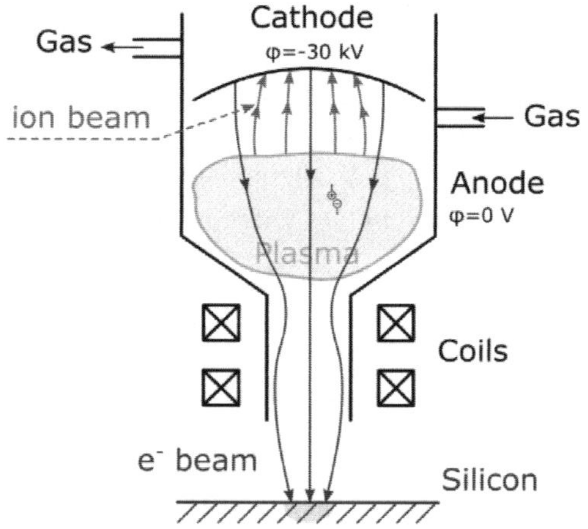

Figure 2. A general scheme of a beam glow discharge based electron gun.

Heating up of the plasma by the electron beam creates a positive back-feed loop in the system and ensures continuing ion-electron pair generation in the plasma. Typically, the plasma is not at equilibrium, as the electron gas temperature is much higher than ion and neutral gas temperature. The electron gas pressure is balanced by the electric field pressure at the plasma interface which determines its form.

Thus, to obtain a quasi-stationary solution for electron beam formation, several physical models should be coupled together. Electric and magnetic field distributions determine particle trajectories. Particle trajectories determine electric charge distribution which affects the electric field. Finally, the plasma interface determines both the particle trajectories and the electric field distribution. Below we describe an iterative algorithm that couples these models.

4. Modeling the electron beam

4.1. Modeling the EM field

The gun geometry is axisymmetric, therefore it was assumed that the electric and magnetic field distributions in the gun are also axisymmetric.

$\varphi = -30\,\text{kV}$ potential is applied to the cathode and its shield, see figure 3. A zero potential $\varphi = 0\,\text{V}$ is set on the gun chamber walls and anode. The gun chamber volume can be divided into three different regions with separate properties. The plasma region is partially ionized gas near the anode, and it has electrical conductivity comparable to that of metals. Therefore, it is assumed that the electric field in the plasma is practically zero. The plasma is considered quasi-neutral (i.e. concentrations of electrons and positive ions in it is approximately the same, $n_e \approx n_p$). Nevertheless due to higher electron mobility, it becomes slightly charged, and the plasma potential φ_{plasma} is slightly positive.

The working gas near the cathode, however, is characterized by very low ionization level and by a very strong electric field ($E \sim 10^6 V/m$) caused by the potential difference between the plasma and walls on the one hand and the cathode on the other hand. This electric field accelerates electrons and positive ions emitted from the cathode and plasma, respectively, so that an electric charge distribution is created in the gas. The charge density is higher near the cathode and plasma surfaces where particle velocities are lower. Thus the electric field in the working gas is modeled by solving a Poisson equation for the electric potential:

$$\Delta\varphi = -\frac{\rho}{\varepsilon_0} \qquad (1)$$

where the distribution of the electric charge ρ is calculated from the particle trajectories.

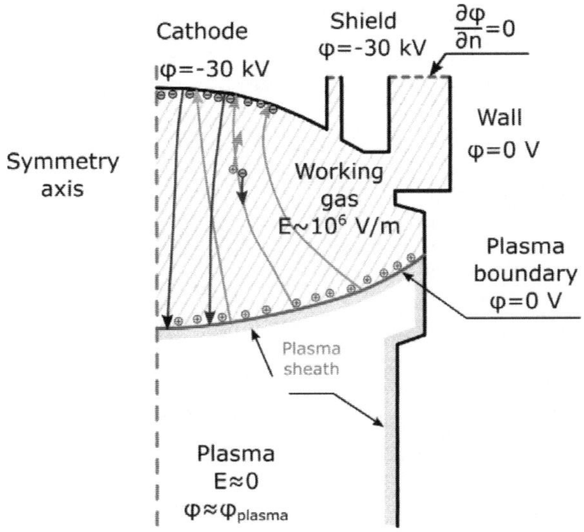

Figure 3. Boundary conditions in electric field simulations.

The plasma and the working gas is separated by a very thin boundary layer called the plasma sheath. It is an electrically charged layer in which the positive ions are accelerated, and electrons are slowed down to prevent their escape from the plasma. Because the boundary between the plasma sheath and the working gas is not clearly defined, we assume for simplicity that it corresponds to the zero electric potential isoline. For this reason we use $\varphi = 0\,\text{V}$ boundary condition on the gas-plasma interface.

The equation for an axisymmetric magnetic field distribution can be split in two separate equations, one for the azimuthal magnetic field component $\vec{H}_\theta$, and one for the azimuthal magnetic vector potential component $\vec{A}_\theta$:

$$\nabla\times\left(\nabla\times\vec{H}_\theta\right) = \nabla\times\vec{j}_{r,z} \qquad \nabla\times\left(\frac{1}{\mu\mu_0}\nabla\times\vec{A}_\theta\right) = \vec{j}_\theta \qquad (2)$$

where $\vec{H}_\theta = H_\theta\vec{e}_\theta$ (and similarly $\vec{A}_\theta$ and $\vec{j}_\theta$), and $\vec{j}_{r,z} = j_r\vec{e}_r + j_z\vec{e}_z$ is a meridional vector.

Although both equations systems can be solved by our model, only $\vec{A}_\theta$ equation is used in practice, because electromagnet current strength (~1000 A-windings) by far exceed the beam current strength (1-3 A). The simulation area for the magnetic field encompasses the gun and a rectangular area around it. Symmetry boundary condition is applied on rotation axis and natural boundary condition on other boundaries.

The material properties of the ferromagnetic shields of the electromagnets could not have been determined. Simulations have shown that a typical magnetic field strength in the shields is about 0.1-0.3 T. It was decided that a constant permeability $\mu = 1000$ would be a good approximation of the shield material. A test simulation with $\mu = 20$ showed no significant changes in the particle impact distribution on the cathode or anode surfaces, which was our primary concern.

4.2. Calculation of plasma interface position

The plasma interface must obtain a shape that ensures that the electron gas pressure at the plasma interface is equal to electric field pressure [5]:

$$p_E = \frac{\varepsilon_0 E_n^2}{2} = n_e k T_e \tag{3}$$

where E_n is the normal component of the electric field at the plasma interface, n_e is electron concentration in the plasma, k is the Boltzmann constant, and T_e is electron temperature. Since we do not model the electron concentration and temperature in these simulations, we simply assume a constant p_E value. Several p_E values were used in the simulations to obtain a better agreement with experiments.

The following iterative algorithm is applied to find the interface position: first, electric field distribution is calculated, then an equation for interface node shifts h_i is solved:

$$h_i + \lambda \frac{\partial^2 h_i}{\partial s^2} = p_E - \frac{\varepsilon_0 E_{n,i}^2}{2} \tag{4}$$

where λ is a smoothing factor and s is interface arc length. The interface nodes are shifted in the normal direction, $\delta \vec{r}_i = h_i \xi \vec{n}_i$, where is ξ some parameter. The node on the symmetry axis is shifted in the vertical direction, and the interface cross point with the wall is found by linear continuation of the penultimate segment of the interface.

4.3. Modeling particle trajectories and charge distribution

A predefined number of ion trajectories is calculated by integrating the Newton's equations

$$m\vec{a} = q(\vec{E} + \vec{v} \times \vec{B}) \tag{5}$$

with 4th order Runge-Kutta method. It is assumed that emitted ions were protons. In [5], the Bohm's speed is used to determine the ion emission current density j_p from the plasma surface:

$$j_p \sim n_p \sqrt{\frac{2kT_e}{m_p}} \tag{6}$$

where m_p is the mass of a positive ion. Assuming a homogeneous distribution of the electron temperature and ion concentration, a homogeneous emission rate for the ions can be used. The initial positions of the ion trajectories are chosen by randomly picking interface meridional segments and positions on them according to surface area represented by the segments. Zero velocity is used as initial condition. Each trajectory corresponds to part of the total ion current:

$$I_p = \frac{I_{ion}}{N} \tag{7}$$

where I_{ion} is predefined ion current strength, and N is the number of the ion trajectories.

Table 1. Secondary electron emission coefficient for H^+ hits on Al surface at 250°C, [6].

Kinetic energy, keV	9.9	20.12	30.2	41.08
κ	0.69	0.98	1.17	1.27

When an ion hits the cathode, an electron trajectory is generated with electron current $I_e = -\kappa I_p$, where κ is the number of emitted secondary electrons. Initially we used κ from [6], table 1, however, the efficiency of the gun (~90%) suggests that κ is much larger, and simulations with $\kappa = 9$ where also performed. Electron trajectories are calculated until they hit any part of the gun elements or boundary of the simulation area.

Table 2. Cross sections for hydrogen ionization, [7]. Values for H_2 are doubled values for H.

Kinetic energy, keV	9.4	11.4	13.4	15.4	18.4	22.4	26.4	32.4
σ for H, Å^2	0.162	0.245	0.331	0.438	0.569	0.779	1.053	1.227
σ for H_2, Å^2	0.324	0.490	0.662	0.876	1.138	1.558	2.106	2.454

We were also concerned with ion-electron pair generation in the neutral gas due to ionization by fast moving ions. For this reason, an ionization effect was implemented. During a time step, an ion travels a distance l. Its kinetic energy determines the reaction cross section σ, table 2, and reaction probability $\Gamma = 1 - e^{-n\sigma l}$, where n is the gas particle concentration calculated from a given gas pressure p. If a reaction occurs, a pair of ion and electron trajectories is generated with currents I_p and $-I_p$.

The electric charge density for a mesh node is calculated by determining time Δt spent by each trajectory in the node neighborhood S. Charges $I_{p,e}\Delta t$ are added to the total node charge, and later it is divided by the 3D volume V obtained by rotating S, see figure 4.

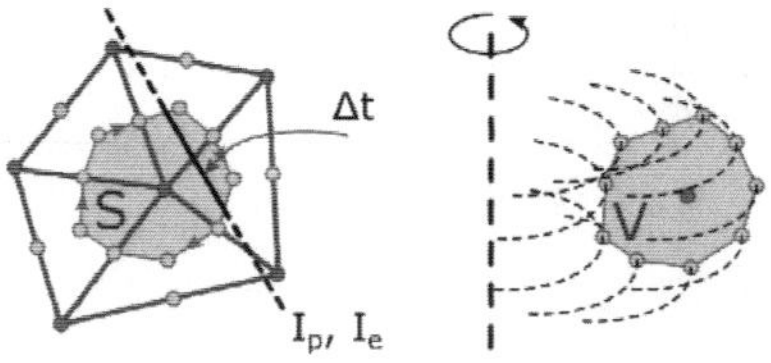

Figure 4. A node neighborhood area S in 2D meridional plane can be rotated around the symmetry axis to obtain the associated 3D charge distribution volume V.

5. Numerical implementation

Triangular meshes for discretization of the electric and magnetic field equations were generated with a mesh generation program *gmsh* [8]. A single mesh for both electric and magnetic fields is created, figure 5. For the electric field simulations, only part of the whole mesh is used.

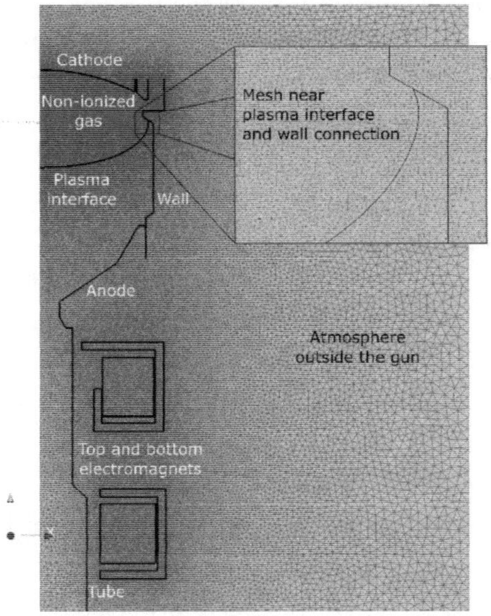

Figure 5. Example of 2D triangular mesh generated by *gmsh*.

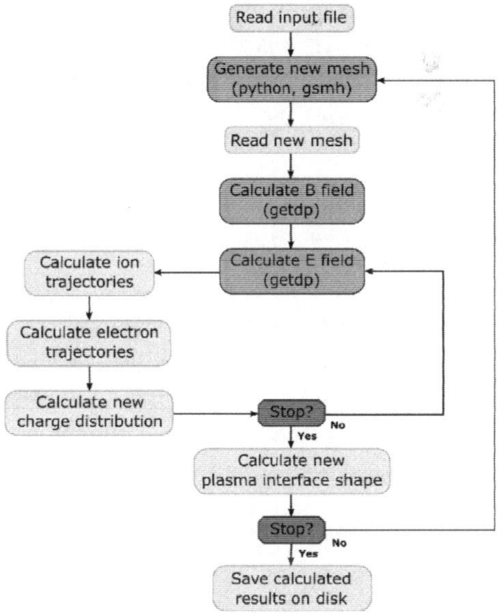

Figure 6. A block scheme of the main iteration cycle for the electron beam simulation.

The equations were discretized using a finite element program *getdp* [9]. Both *gmsh* and *getdp* use special text input files, which are generated by a special *python* language script, which also calls *gmsh* to generate the mesh.

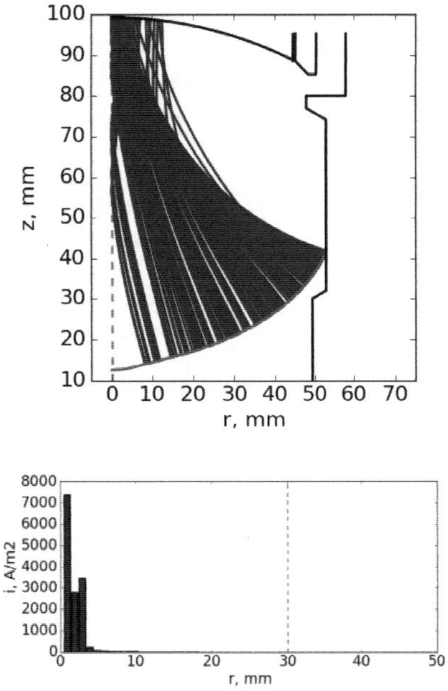

Figure 7. Calculated ion particle trajectories (top) and impact density distribution on the cathode (bottom). $p_E = 0.01\,\text{Pa}$, $p = 1.5\,\text{Pa}$, $I_{ion} = 0.1\,\text{A}$.

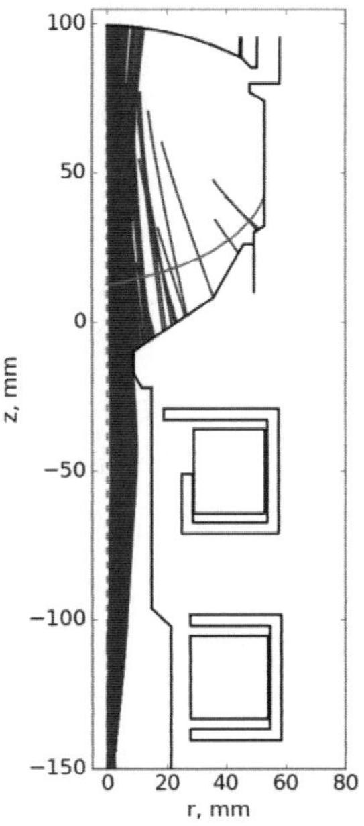

Figure 8. Calculated electron trajectories. $p_E = 0.01\,\text{Pa}$, $p = 1.5\,\text{Pa}$, $I_{ion} = 0.1\,\text{A}$.

The main iteration cycle as well as the particle trajectory and plasma interface models are implemented in the main program written in *C++*. The block scheme of the main iteration cycle that couples all models together can be seen in figure 6. The program reads its input parameters from a text file and can be run in two modes: the main mode and preparation mode, which allows to calculate an initial plasma interface position by iteratively calculating the electric field distribution without particle trajectories.

6. Simulation results

Considering large secondary electron emission, a small ion current strength was chosen (0.1 A). Due to uncertainty about the actual electron gas concentration and temperature, several different electron gas pressure values were tested, form very low ($p_E = 0.01$ Pa) to very high ($p_E = 1.8$ Pa).

The calculation results with $p_E = 0.01$ Pa showed a very small ion impact area on the cathode (~8 mm in diameter), and a narrow electron beam, see figures 7 and 8. However, experimental observations, figure 9, show that impact area can be as large as 60 mm in diameter.

Simulations with larger electron gas pressure have showed that the calculated plasma interface shifts upwards when p_E is increased, see figure 10. As a result, the ion emission spreads over larger cathode area. The best agreement with the experimental results is obtained for $p_E = 1.8$ Pa, see figure 11.

A significant part of electrons hits the anode. Measured by the spike in the electron impact distribution, see figures 12 and 13, the affected area has diameter between 30 and 40 mm, which approximately corresponds to dimensions observed in experiments (~30-35 mm).

Figure 9. Example of a cathode surface that have been bombarded by ions. The impact area diameter is ~60 mm.

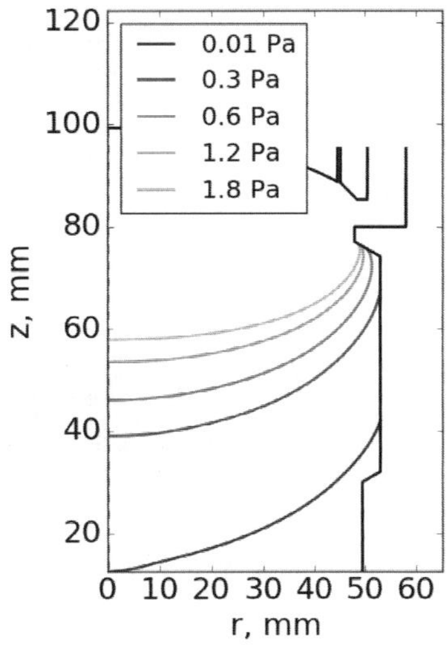

Figure 10. Dependence of the plasma interface shape on the applied electron gas pressure. $p = 1.5\,\text{Pa}$, $I_{ion} = 0.1\,\text{A}$.

Due to electron current losses on the anode, the calculate gun efficiency rate is very low (~0.24 for $p_E = 0.6\,\text{Pa}$ and ~0.13 for $p_E = 1.8\,\text{Pa}$. This suggests that there might be an inhomogeneous ion emission distribution on the plasma interface, in which the emission maximum is centered on the symmetry axis. Such an emission distribution could be caused by the interaction between the plasma and the electron beam. This effect could be modeled by using, for example, bell-shaped ion emission distributions instead of homogeneous emission rates. The electron gas pressure distribution on the plasma interface also can be modeled as a bell-shaped function. Due strong plasma shape dependence on the electron gas pressure, there is also a need to relate it to the applied working gas pressure and generated beam current strength.

7. Conclusions

A 2D quasi-stationary model for electron beam formation in a beam glow discharge based electron gun has been developed and tested. The model includes 2D axisymmetric steady-state modules for electric and magnetic field simulations and for calculation of 3D ion and electron trajectories.

Simulations have shown that the model is able to model the basic properties of the electron beam, such as dimensions of particle impact areas on the cathode and anode. However, further improvements have to be introduced in order to model correctly electron trajectory distribution in the beam, and the beam current dependence on the applied gas pressure.

8. References

[1] Becker R and Herrmannsfeldt W B 2004 *J. Phys.: Conf. Series* **2** 152-163
[2] OPERA 3d, *Cobham Technical Services, Vector Fields Software,* Oxford, UK
[3] Elliott S M and Delferrière O and Simkin J and White E K 2008 *Proc. ECRIS08 (Chicago)* p 220
[4] Read F H and Bowring N J 2004 *Nuclear Instruments and Methods in Phys. Research A* **531** 407-415
[5] Melnik I V 2005 *Radioelektronika (in Russian)* **6** 61-71
[6] Svensson B and Holmèn G 1982 *Phys. Rev. B* **25** 3056–62
[7] Shah M B, Elliot D S and Gilbody H B 1987 *J. Phys. B: At. Mol. Phys.* **20** 2481–85
[8] Geuzaine C and Remacle J-F 2009 *Int. J. for Num. Methods in Engineering* **79** 1309–31

[9] Geuzaine C 2007 GetDP: a general finite-element solver for the Rham complex *Sixth International Congress on Industrial Applied Mathematics (ICIAM07) and GAMM Annual Meeting, PAMM* vol 7 (Zuerich: Wiley) p 1010603–4

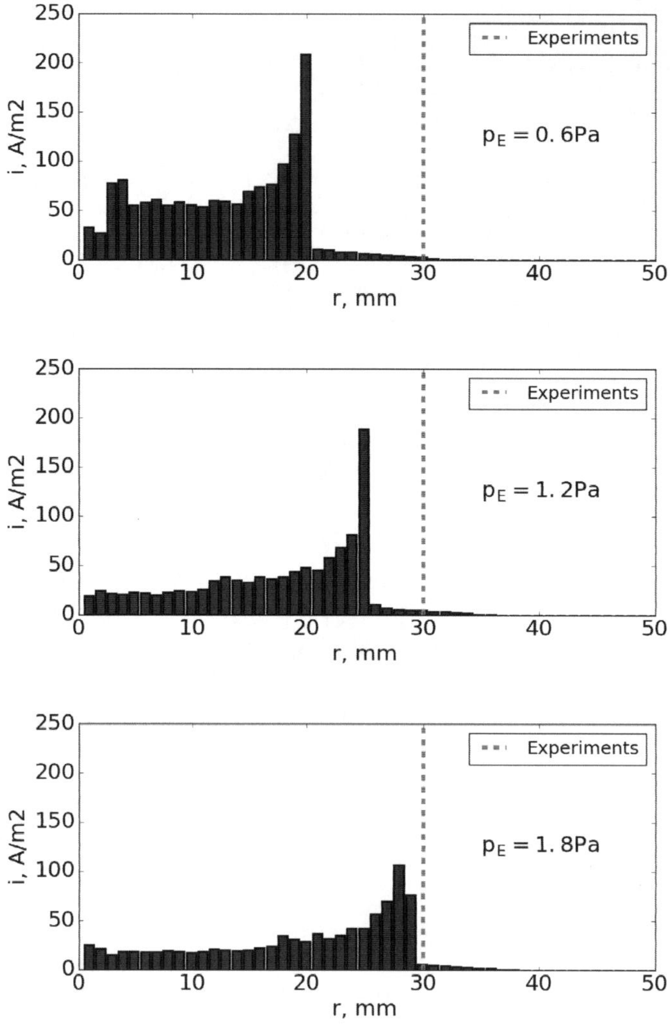

Figure 11. Dependence of the radial distribution of ion impact density on the cathode on the applied electron gas pressure. $p = 1.5\,\text{Pa}$, $I_{ion} = 0.1\,\text{A}$.

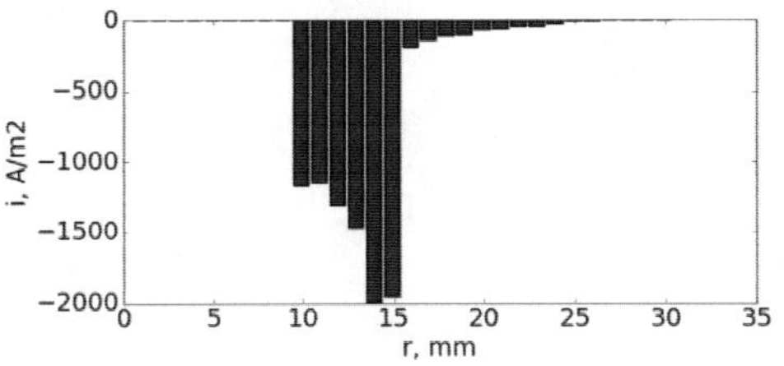

Figure 12. Radial distribution of electron impacts on the anode for $p_E = 0.6$ Pa, $p = 1.5$ Pa, $I_{ion} = 0.1$ A.

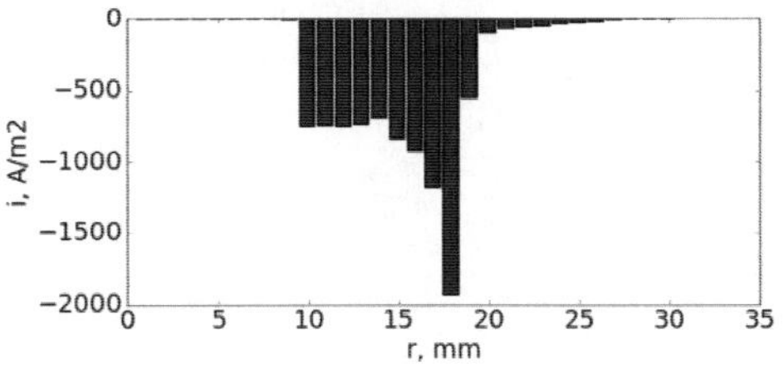

Figure 13. Radial distribution of electron impacts on the anode for $p_E = 1.8$ Pa, $p = 1.5$ Pa, $I_{ion} = 0.1$ A.

Acknowledgments
The research leading to these results has been supported by the European Regional Development Fund project "Development of Competence Centre of Mechanical Engineering", contract No.1.2.1.1/16/A/003 signed between Competence Centre of Mechanical Engineering and Central Finance and Contracting Agency, Research No. 2.3 "Modelling and optimization of plasma shape in electron gun heater for the improvement of crystal growth process control".

IOP Conf. Series: Materials Science and Engineering **355** (2018) 012009 doi:10.1088/1757-899X/355/1/012009

Analysis of the impact of modification of cold crucible design on the efficiency of the cold crucible induction furnace

R Przylucki[1], S Golak[1], P Bulinski[2], J Smolka[2], M Palacz[2], G Siwiec[3], J Lipart[3] and L Blacha[3]

[1]Department of Industrial Informatics, Silesian University of Technology, Krasinskiego 8, 40-019 Katowice, Poland
[2] Institute of Thermal Technology Silesian, University of Technology, Konarskiego 22, 44-100 Gliwice, Poland
[3]Institute of Metals Technology, Silesian University of Technology, Krasinskiego 8, 40-019 Katowice, Poland

E-mail: roman.przylucki@polsl.pl

Abstract. The article includes numerical simulation results for two induction furnace with cold crucible (IFCC). Induction furnaces differ in cold crucible design, while the inductor geometry was preserved for both variants. Numerical simulations were conducted as three dimensional one, with coupled analysis of electromagnetic, thermal and fluid dynamics fields. During the experiment, six calculation variants, differ in amount of molten titanium (three different weights of titanium for each type of cold crucible) were considered. Main parameters controlled during the calculations were: electrical efficiency of the IFCC and the meniscus shape of liquid metal.

1. Introduction

Induction furnaces with cold crucible are most commonly used for melting very reactive metals and alloys. Such metals should be melted without contact with the crucible. Cold crucible is a compromise solution. The best, from the point of view of lack of contact, is the levitation melting. But in levitation melting there are problems with the lifting force especially for heavy workpieces. In cold crucible furnace, the molten metal contacts only with a bottom of the crucible. From the crucible walls, liquid metal is repulsed by electrodynamic forces. At the bottom, the metal is in contact with the crucible not directly, but through a solidified layer (skull). There are two main problems in cold crucible furnaces design: electric (and of course total) efficiency, and minimization of the skull [1].

Low electric efficiency is due to double conversion of electrical energy: from current to the electromagnetic field and again to current. First conversion between inductor and cold crucible segment (finger) and the second between cold crucible segments and the molten metal. This cannot be changed but energy losses can be reduced by the appropriate cold crucible construction [2]. Moreover, the skull volume depends on the cold crucible construction.

The aim of the article is the modification of the design of cold crucible to improve the efficiency of the melting process and to minimize the skull.

Content from this work may be used under the terms of the Creative Commons Attribution 3.0 licence. Any further distribution of this work must maintain attribution to the author(s) and the title of the work, journal citation and DOI.
Published under licence by IOP Publishing Ltd

2. Models of considered cold crucibles

Two cold crucible designs are investigated in this article. Simplified sketch of the first and the second construction is presented in Figure 1 and 2, respectively. As shown in Figure 1 first construction was made from one piece of copper. There is a separation between segments, and the incision also includes part of the crucible bottom (orange part with index 6). There is a massive chassis below the inductor. The shuns does not have pole pieces. The crucible bottom is slightly concave. Advantage of this design is the incision of the outer part of bottom which prevents shorting of the currents induced in the bottom part of the segments. Disadvantages of this construction are: massive chassis just below the inductor, and lack of pole pieces. Problematic is also the recess in the bottom. This causes a thick skull.

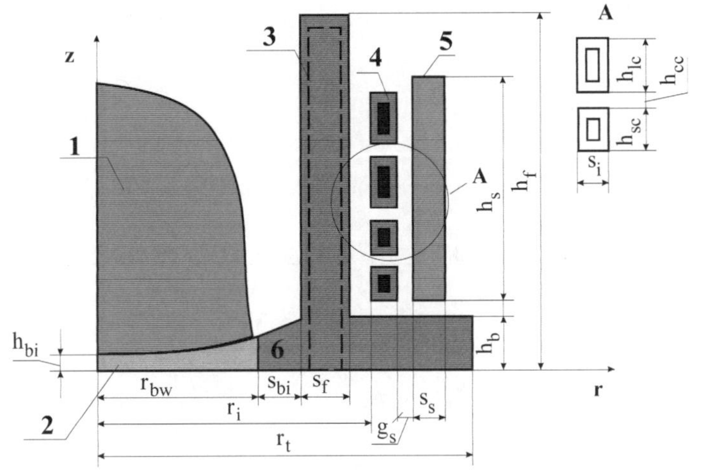

Figure 1. Sketch of first IFCC Model, 1 – metal melt, 2 – solid bottom, 3 – segment, 4 – inductor, 5 – shunt, 6 – separated part of the bottom.

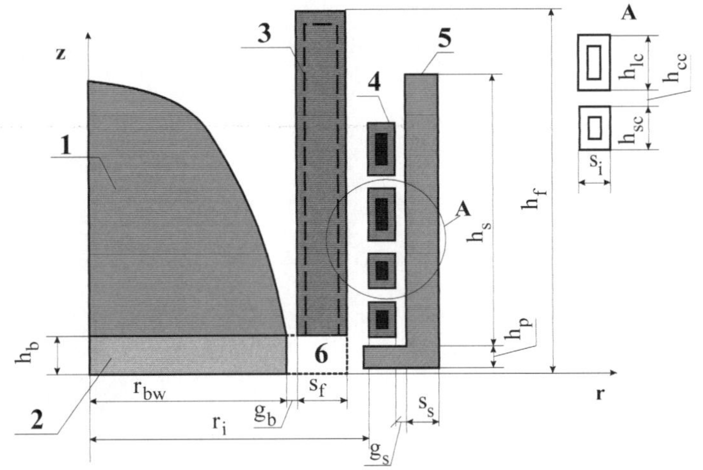

Figure 2. Sketch of second IFCC Model, 1 – metal melt, 2 – solid bottom, 3 – segment, 4 – inductor, 5 – shunt, 6 – ceramic element.

The second construction of the cold crucible, considered in this article, consists of two main, separate pieces: segments and flat bottom. The shunts has pole pieces below the inductor, and there is no conducting material just below inductor and segments. Such a design (more difficult from the construction point of view) prevents the induction of currents outside the cold crucible segments.

2.1. Model for electromagnetic field analysis

Analysis of electromagnetic field basis on A-V formulation equation (1) [3] is conducted as a steady state one. Electromagnetic field analysis results were volumetric heat sources equation (2) [4] and volumetric electromagnetic forces equation (4) [4]. These values were calculated in the melt area only, and they are input data for temperature and fluid dynamics analysis.

$$\mathrm{rot}\left(\frac{1}{\mu_0}\frac{1}{\mu}\mathrm{rot}\,\mathbf{A}\right)+\boldsymbol{\sigma}\left(j\omega\mathbf{A}+\mathrm{grad}\,V\right)=0 \tag{1}$$

where:

μ_0 - permeability of vacuum; μ - relative permeability; **A** - magnetic vector potential; σ - conductivity; ω - angular velocity; V - scalar electric potential; j - unit imaginary number.

$$q_v=\frac{1}{2}|\mathbf{J}|^2\cdot\frac{1}{\sigma} \tag{2}$$

where: q_v - volumetric density of active power; **J** - current density.

$$\mathbf{J}=-\sigma\left(j\omega\mathbf{A}+\mathrm{grad}\,V\right) \tag{3}$$

$$\mathbf{f}_m=\frac{1}{2}\Re\mathrm{e}\,(\mathbf{J}\times\mathbf{B}^*) \tag{4}$$

$$\mathbf{B}=\mathrm{rot}\,\mathbf{A} \tag{5}$$

where:

$\mathbf{f}_m$ - time average for period volumetric force density; **B*** - conjugate value of induction.

The problem was considered as coupled analysis of: electromagnetic, thermal and fluid dynamics fields in 3-D domain. Electromagnetic field analysis was performed with use of getDP program [5]. For the temperature and fluid dynamics analysis, Ansys-Fluent program was used.

For electromagnetic field calculation, only 1/16 of the device was modelled (Figure 3). It was possible because of periodicity of the device construction.

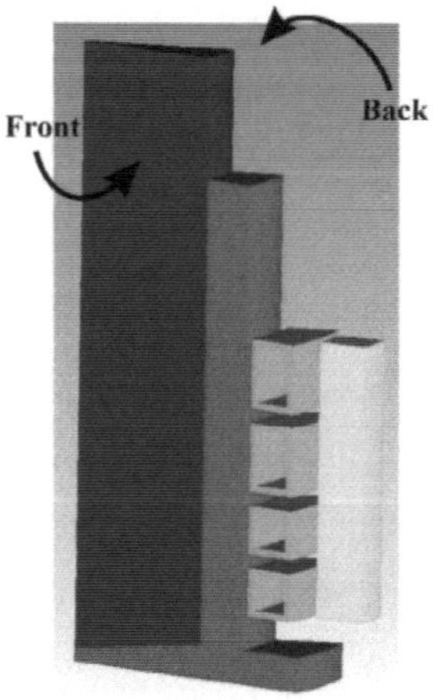

Figure 3. Calculation model for electromagnetic field analysis of the first IFCC.

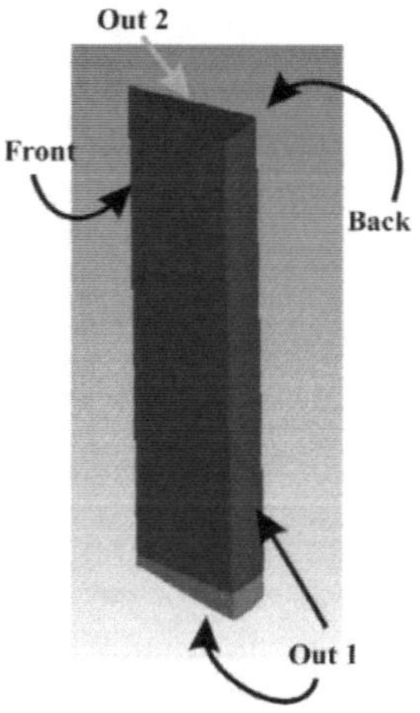

Figure 4. Calculation model for thermal and fluid dynamics field analysis of the first IFCC.

On front and back model surfaces (Figure 3), the periodicity boundary condition was applied. In other directions, the computational model is surrounded by air and on the outside air surfaces boundary Dirichlet condition setting potentials to zero was applied.

Basic dimensions of the models are presented in Table 1. Bottom and segments of crucible were made of copper of the conductivity of $56 \cdot 10^6$ S/m, melt metal was made of titanium with the conductivity of $50.5 \cdot 10^4$ S/m. Shunts were made of Fluxtrol 100.

Table 1. Model parameters.

Parameter (mm)	Model 1	Model 2
r_{bw}	40	45
s_{bi}	6	1
s_f	15	15
g_s	2	2
s_s	10	10
s_i	13	13
r_i	75	75
r_t	90	-
h_{bi}	4	10
h_b	10	10
h_s	85	85
h_p	-	7.5
h_f	144	144
h_{sc}	3	13
h_{lc}	25	25
h_{cc}	4	4

2.2. *Model for temperature field and fluid dynamics analysis*

Analysis of temperature field and fluid dynamics were performed in Ansys Fluent program. The calculation domain was limited to the melt area and a bottom of crucible only, as presented in Figure 4. Temperature calculations were based on equation (6) [6].

$$\frac{\partial}{\partial t}(\rho c T) + \rho c \mathbf{v} \operatorname{grad} T = \operatorname{div}(k \operatorname{grad} T) + q_v + q_r \tag{6}$$

where:

ρ - density; c - specific heat; T - temperature; **v** – velocity vector; k - thermal conductivity; q_v - volumetric heat source from electromagnetic calculation; q_r - volumetric density of radiative heat exchange derived on the basis of Gauss-Ostrogradsky theorem from surface radiative heat transfer [7].

As in the case of electromagnetic field analysis, the calculation domain was limited to 1/16 of the device (Figure 4). On the front and back model surfaces adiabatic boundary condition was applied, on the outside surface in the crucible segments and below bottom (out 1, Figure 4.) Dirichlet boundary condition setting temperature to 303 K was applied, on the free molten metal surface (out 2), the heat loss with convection and radiation were taken into consideration.

The fluid dynamics calculations should determine the shape of free surface of molten metal in the IFCC and therefore the flow field within crucible was treated as multiphase one. The Volume of Fluid approach was applied in the calculation model. To obtain the flow field distribution, the continuity equation in the form equation (7) [6] and momentum conservation equation (8) [6] should be solved.

$$\frac{\partial}{\partial t}\left(\alpha_q \rho_q\right)+\operatorname{div}\left(\alpha_q \rho_q \mathbf{v}\right) \tag{7}$$

where: α_q - volume fraction of the qth phase; ρ_q - density of the qth phase.

$$\frac{\partial}{\partial t}(\rho\mathbf{v})+\operatorname{div}(\rho\mathbf{v}\mathbf{v})=-\operatorname{grad} p+\varepsilon(\operatorname{grad}\operatorname{div}\mathbf{v}-\operatorname{rot}\operatorname{rot}\mathbf{v})+\rho\mathbf{g}+f_m+f_s \tag{8}$$

where: p - pressure; ε - dynamic viscosity; **g** - gravitational acceleration; f_s - volume force density calculated from surface tension derived on the basis of Gauss-Ostrogradski theorem [8].

These equations are supplemented with boundary condition, on the front and back surfaces periodicity boundary condition was assumed. On the top of model pressure outlet condition was adopted, for the free surface of molten metal shear stress equal to 0 on both sides was assumed. On the outside wall of the model (out 2), no slip condition was assumed.

2.3. Coupling procedure

Electromagnetic, heat transfer and fluid flow calculation were conducted for two different numerical submodels. Models differs in geometry and calculation domain. Calculation model for electromagnetic field consists of molten metal, crucible, inductor, shunt and air. Temperature and fluid dynamic calculations were conducted for the model limited to the molten metal and air above free surface of metal, and bottom of the crucible.

Fluid dynamics and temperature calculations were performed in time domain, and the time step was 0.00001 s. Each 25 time steps the calculations of electromagnetic field were performed and mean for the period values of forces and power sources were determined, and then transferred as the source terms for the thermal submodel. Before the transfer, the forces and heat sources from electromagnetic calculation were approximated for the thermal - fluid dynamic calculation mesh [9].

3. Numerical experiment

Six simulation variants were performed during the experiment. For each of the two IFCC models, three calculation variants, different in volume of the titanium, were conducted. The variant designators are as follows: IFCC model 1, volume of titanium 1 is designated as m1v1. Considered quantities of titanium are: 1.06 kg, 1.6 kg, 2.13 kg which corresponds to complete filling the crucible to the level of 0.04 m (27%), 0.06 m (42%), 0.08 m (55%).

All variants of calculations were conducted for the same inductor current and current frequency of 10 kHz. The liquid metal properties were as follows: conductivity [10] $\sigma = 50.5\cdot10^4$ S/m density [11] $\rho = 4160$ kg/m^3, thermal conductivity [12] $\lambda= 32$ W/(m·K), specific heat $c = 9.87\cdot10^2$ J/(kg·K), dynamic viscosity [13] $\varepsilon = 4.03\cdot10^{-3}$ (Pa·s).

Due to the difficulty of coupling of the model to the electromagnetic field analysis with the model for temperature and FD field analysis, the calculations were made only partially. In the Figure 5, and Figure 6 calculation results for the m1v1 variant for two different supply powers 20% of the maximum power (40 kW) and 40% of the maximum power (80 kW) are shown. Figures 5 and 6 shows meniscus shape. As can be seen, for 20% of the power the liquid metal is not completely pushed away from the walls of the crucible, but for the 40% of the power the liquid metal is completely pushed away from the wall of the crucible practically from the bottom of the IFCC. But at this stage of the calculations, no metal solidification has yet be taken into account.

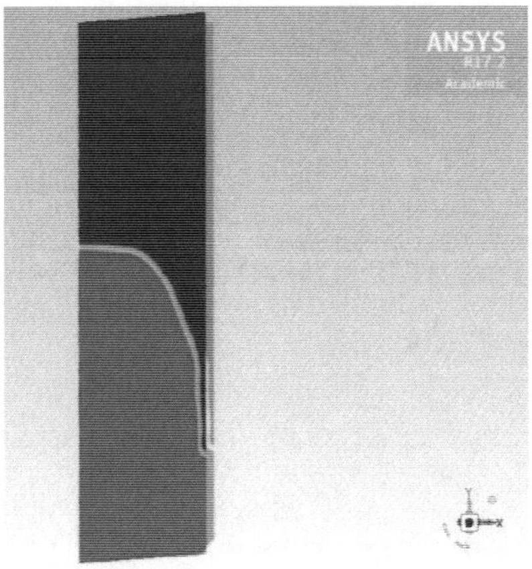

Figure 5. Metal meniscus shape for 20% of the supply power.

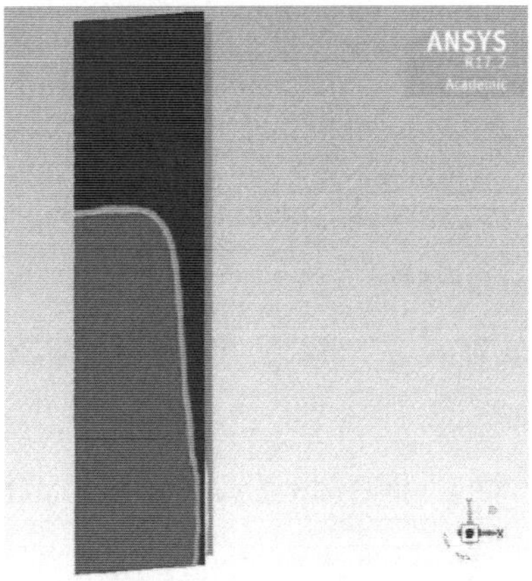

Figure 6. Metal meniscus shape for 40% of the supply power.

References

[1] Spitans S, Baake E, Jakovics A and Franz H 2015 Numerical simulation of electromagnetic levitation in cold crucible furnace *Magnetohydrodynamics* **51** 567-578

[2] Rot D, Jirinec S, Kozeny J and Poznyak I 2015 Electrical efficiency of induction furnace with cold crucible via different segments width *Electric Power Engineering (EPE) 16th International Conference* IEE DOI 10.1109/EPE.2015.7161132

[3] Fort J, Garnich M and Klymyshyn N 2005 Electromagnetic and thermal-flow modelling of a cold-wall crucible induction melter *Metallurgical and Materials Transaction B* **36** 141-152

[4] Flux 2005 *3D Vol5 Complement on the formulations 3D* Cedrat 138-171.

[5] Dular P, Geuzaine C *GetDP reference manual: the documentation for GetDP, a general environmental for the treatment of discrete problems* getdp.info

[6] *Ansys 15.0 manual* www.ansys.com

[7] *Fluent 6.3 User's Guide 13.2.1 Heat transfer theory* https://www.sharcnet.ca/Software/Fluent6/html/ug/node568.htm

[8] *Fluent 6.3 User's Guide 23.3.8 Surface Tension and Wall Adhesion* https://www.sharcnet.ca/Software/Fluent6/html/ug/node888.htm

[9] Bulinski P, Smolka J, Golak S, Przylucki R, Palacz M, Siwiec G, Lipart J, Bialecki R and Blacha L 2017 Numerical and experimental investigation of heat transfer process in electromagnetically driven flow within vacuum induction furnace *Applied Thermal Engineering* **124** 1003-13

[10] Seydel U and Fucke W 1980 Electrical resistivity of liquid Ti, V, Mo and W *Journal of Physics F: Metal Physics* **10**

[11] Ishikawa T 2005 Thermophysical properties of molten refractory metals measured by an elecrostatic levitator *Journal of Electric Materials*

[12] Mills K, Monaghan B and Knee J 1996 Thermal conductivities of liquid metals *Proceedings of the Twenty-Third International Thermal Conductivity Conference*

[13] Pardis P, Ishikawa T and Yoda S 2002 Non contact measurements of surface tensions and viscosity of niobium, zirconium and titanium using and electrostatic levitation furnace *International Journal of Thermophysics* **23**

Acknowledgments

Financial assistance was provided by grant no. 2014/13/B/ST8/02364 funded by the National Science Centre, Poland, and is here acknowledged.

IOP Conf. Series: Materials Science and Engineering **355** (2018) 012010 doi:10.1088/1757-899X/355/1/012010

A feasibility study for high-temperature titanium reduction from $TiCl_4$ using a magnesiothermic process

S L Ivanov and D Zablotsky

Institute of Physics, University of Latvia (IPUL), Miera str. 32, Salaspils, LV-2169, Latvia

E-mail: ivanov@sal.lv

Abstract. The current industrial practice for titanium extraction is a complex procedure, which produces a porous reaction mass of sintered titanium particulates fused to a steel retort wall with magnesium and $MgCl_2$ trapped in the interstices. The reactor temperature is limited to approx. 900 °C due to the formation of fusible TiFe eutectic, which corrodes the retort and degrades the quality of titanium sponge. Here we examine the theoretical foundations and technological possibilities to design a shielded retort of niobium-zirconium alloy NbZr(1%), which is resistant to corrosion by titanium at high temperature. We consider the reactor at a temperature of approx. 1150 °C. Supplying stoichiometric quantities of reagents enables the reaction in the gas phase, whereas the exothermic process sustains the combustion of the reaction zone. When the pathway to the condenser is open, vacuum separation and evacuation of vaporized magnesium dichloride and excess magnesium into the water-cooled condenser take place. As both the reaction and the evacuation occur within seconds, the yield of the extraction is improved. We anticipate new possibilities for designing a device combining the retort function to conduct the reduction in the gas phase with fast vacuum separation of the reaction products and distillation of magnesium dichloride.

1. Introduction

The primary industrial process for producing metallic titanium is conducted by chlorination of enriched titanium ore and magnesiothermic reduction of titanium tetrachloride $TiCl_4$ in a steel vessel [1]. Under high temperature iron and nickel contained in the reactor walls and titanium form a liquid eutectic [2,3]. The erosion by titanium leads to a structural weakness of the vessel and contamination of the reaction mass, whereby 50-90% of the titanium in the vicinity of the walls is classified as ferrotitanium or scrap. Hence, the temperature of the process is limited to approx. 900 °C. The reduction of $TiCl_4$ proceeds via intermediate stages with formation of lower titanium chlorides. The system $TiCl_4$-$TiCl_3$-$TiCl_2$-TiCl-Ti-Mg-MgCl-$MgCl_2$ permits about 70 potential reaction pathways, the major ones (Figure 1) are determined by thermodynamics and reaction kinetics [4]. The first most probable phase comprises the reduction of vaporized $TiCl_4$ by magnesium to $TiCl_2$ and further as it requires the participation of just 2 molecules of initial reagents. The formation of lower chlorides is also permitted by way of secondary reactions of chlorides with Ti. The efficiency of the reduction hinges on the kinetics of multicomponent reactions connected by complex pathways in a multiphase environment, whereas the transport of reagents into and out of the reaction zones can be a limiting factor. Theoretical studies [5] show that the temperature range approx. 1200 – 1400 °C is more beneficial for the reduction to metallic titanium. Hence, the improvement of the magnesiothermic process requires a technological outline, whereby the titanium tetrachloride and magnesium would

Content from this work may be used under the terms of the Creative Commons Attribution 3.0 licence. Any further distribution of this work must maintain attribution to the author(s) and the title of the work, journal citation and DOI.
Published under licence by IOP Publishing Ltd

interact at a higher temperature than in the current process resulting in a more favourable kinetics, whereas the alternative - electrochemical route [19] - is severely handicapped by high deposition potential of Ti, multivalence and insolubility of $TiCl_4$ in fused salts with associated technical difficulties.

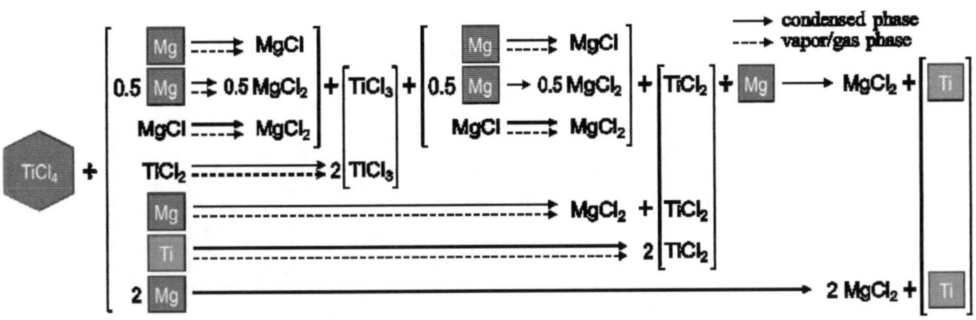

Figure 1. Favorable reaction pathways according to thermodynamics and reaction kinetics (solid lines – condensed phase, dashed – vapor phase).

2. New reactor

The use of niobium alloys offers a considerable increase of the operating temperature and enables rapid disposal of magnesium chlorides, which comprise more than 90% of the reaction products, by discharging them into a condenser (see also Figure 3, left). In the past the IPUL has accumulated experience with NbZr lithium test loops at high temperature (>1000 °C) under vacuum [6]. It is known that niobium is resistant to corrosion by magnesium vapor up to approx. 1150 °C [7]. Preliminary experiments [8] on magnesiothermic reduction of titanium tetrachloride up to 1100 °C in a test reactor made of NbZr alloy showed no corrosive degradation. This alloy is weldable and therefore suitable for constructing a reaction retort for magnesiothermic reduction in the temperature range 1200-1400 °C. Herein, we explore the theoretical benefits of a high-temperature 1150 °C magnesiothermic route and the potential technological framework to ensure dosed supply of reagents and withdrawal of by-products using fast switching valves to conduct the process predominantly in the gas phase [9, 10, 11]. The high chemical activity of titanium requires conducting all operations under inert gas or in vacuum. The conditions of the reaction are selected to ensure that the reduction proceeds in the gas phase to obtain high-quality metallic titanium as fine powder [12] and that the by-product magnesium dichloride is mostly vaporized. This will reduce the amount of time required to evacuate the reactor by distillation of $MgCl_2$ into the condenser.

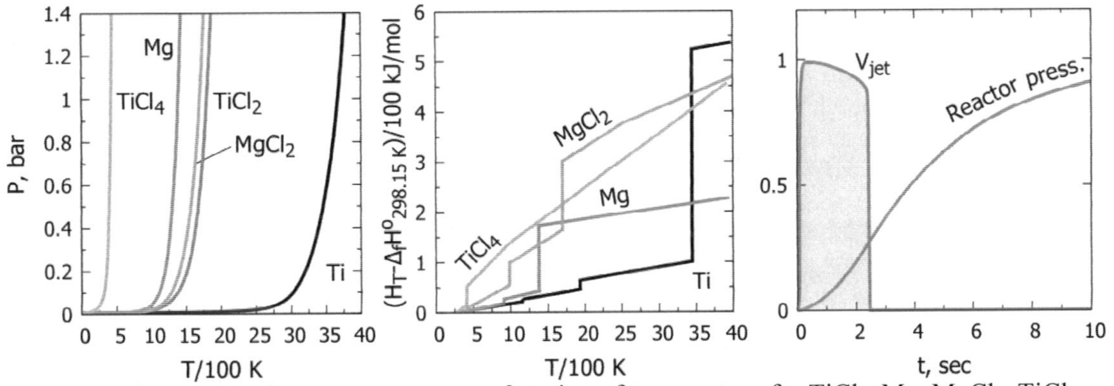

Figure 2. Left - saturated vapor pressure as a function of temperature for $TiCl_4$, Mg, $MgCl_2$, $TiCl_2$ and titanium; middle - enthalpy $\left(H_T - \Delta_f H^{o}_{298.15\,K}\right)$ of 1M of reagents and products of magnesium-

thermal reduction, right - the change of the velocity in the nozzle and of the pressure in the reactor during the supply of magnesium.

The saturated vapors of $MgCl_2$ at 1150 °C in a 10L lab-scale experimental reactor would produce a pressure of 0.137 bar, whereas their total mass is 1.3g (0.0137 M). A 1.3g of $TiCl_4$ and 0.333g of magnesium are supplied to produce this quantity of $MgCl_2$ and 0.327 grams of metallic titanium in accordance with the generalized thermochemical reaction (Figure 1):

$$^1/_2\,TiCl_4 + Mg \rightleftarrows {}^1/_2\,Ti + MgCl_2 + \Delta H_T \tag{1}$$

Hess's law states that the eventual change of enthalpy is independent of the reaction pathway. Hence, the standard enthalpy of a generic reaction is obtainable from the equation

$$\Delta_r H^o_{298.15\,K} = \sum_j \nu_j\, \Delta_f H^o_{298.15\,K}(B_j) - \sum_i \nu_i\, \Delta_f H^o_{298.15\,K}(A_i) \tag{2}$$

where $\Delta_f H^o_{298.15\,K}(X_i)$ is the standard enthalpy of formation [13] for the products (B_i) and initial reagents (A_i), and ν_i are the corresponding molar quantities. Since $\Delta H \gg V\Delta p$ to a sufficient accuracy (approx. ±1%) $\Delta U \approx \Delta H$. Per Kirchhoff's law the enthalpy H_T for reagent Z at temperature T

$$\left(H_T - \Delta_f H^o_{298.15\,K}\right)_Z = \int_{298.15\,K}^{T} C_{p,Z}(t)dt + \sum_\alpha H_{\alpha,Z} \cdot \mathrm{X}(T - T_{\alpha,Z}) \tag{3}$$

where $C_p = \left.\frac{\partial H}{\partial T}\right|_p$ – heat capacity at constant pressure, H_α and T_α are latent heat and temperature for the αth phase change; $\mathrm{X}(x)$ is the Heaviside's step function. The thermal effect of the reaction (1) at temperature T, when the initial reagents (A_i) are supplied at T_{TiCl_4} and T_{Mg} (T_i) is obtainable from the equation

$$\Delta H_T(T_{TiCl_4}, T_{Mg}, T) = \Delta_r H^o_{298.15\,K} + \sum_j \nu_j \left(H_T - \Delta_f H^o_{298.15\,K}\right)_{B_j} - \sum_i \nu_i \left(H_{T_i} - \Delta_f H^o_{298.15\,K}\right)_{A_i} \tag{4}$$

The enthalpies of the reactants and reaction products are shown in Figure 2, middle. The standard enthalpy of the reduction reaction is -239 kJ/mol (per mole of Mg) [4,14]. At 1150 °C the heat of reaction is 258.5 kJ per mole of Mg or 3.54 kJ for 0.0137 M. Assuming the magnesium is supplied at 800 °C, a heat of 3 kJ needed for its evaporation will cool the 10-kg walls by just 0.3 °C. At 1150 °C the vaporized magnesium is reacted with $TiCl_4$, which is injected at 25 °C. Under adiabatic conditions in the reaction zone the reaction products are heated to 2500 °C: magnesium dichloride is superheated vapor, whereas titanium is molten, resulting in a local fusion of titanium particulates. The growing pressure of $MgCl_2$ and $TiCl_2$ vapors facilitates the mass transport out of the reaction zone toward slightly cooler walls entraining the finely dispersed titanium particles and depositing them on the walls. Hence, the presence of titanium in the discharged condensate would testify of the insufficiency of the reaction stage. At the same time the heat of the reaction is insufficient to change the temperature of the massive reactor walls. Most of this heat is removed from the reactor during the distillation of magnesium dichloride. Hence, the overheating of the reactor and the difficulties associated with intense cooling of the retort as in the current industrial practice can be mitigated. Following vaporization of 0.0137 M of magnesium the pressure in the reactor reaches 0.163 bar. After completion of the reduction reaction the vapor pressure of $MgCl_2$ is 0.137 bar. The distillation of the vapors into the condenser at 720 °C reduces the pressure to 43 Pa. Supplying the magnesium in excess of the stoichiometric ratio, the surplus will be discharged along with the by-product $MgCl_2$.

2.1. Injection and vaporization of magnesium

A stoichiometric quantity m_{Mg} of molten magnesium can be injected hydraulically into the reactor from a supplying tank at approx. 800 °C via a magnesium feed line and a nozzle with a small aperture σ_N by opening a controlled heat resistant valve for a certain period of time. The ejection velocity v follows from the unsteady Bernoulli equation

$$\frac{dv}{dt} = \frac{1}{\rho h}\left[\Delta p - 0.5\left(1 + \zeta_N(Re)\right)\rho v^2\right] \tag{5}$$

where $\Delta p = \rho g h - p_V + P$ is the pressure difference at the nozzle comprising hydrostatic contribution, feed tank overpressure (P) and reactor pressure (p_V) (including magnesium surface tension); ζ_N is the nozzle hydraulic resistance coefficient [15], $Re = \nu^{-1} v \sigma_N$ – Reynolds number. The linear stability analysis [16] shows that at a sufficiently small aperture diameter σ_N the liquid jet is susceptible to a Plateau-Rayleigh instability leading to its breakup into an ensemble of liquid droplets beyond a critical length $\sim 4.2\sigma_N\sqrt{We}$ [17] ($We = \rho v^2 \sigma_N \gamma$ – Weber number, γ – liquid magnesium surface tension at approx. 800 °C). Estimates show that the mass loss of a single droplet due to a radiative heat transfer from the walls of the reactor is small (approx. 6%), hence, the necessary conditions for vaporization arise solely at the wall-droplet interface. In the worst-case scenario the heat transfer becomes impeded by an insulating vapor layer due to the Leidenfrost effect. The dynamics of vaporization (and pressure) of spheroidal droplets [18] in this case can be modeled by equation

$$\ln[V(t)V_0^{-1}] = -(\alpha_k + \alpha_r)\frac{T_{wall}-T_{sat}^*}{\delta\rho[r+c(T_{sat}^*-\langle T\rangle)]}t, \quad \alpha_k = \eta \cdot \sqrt{\frac{[r+c\cdot(T_{sat}^*-\langle T\rangle)]}{(T_{wall}-T_{sat}^*)}\lambda^* D^{-1}} \cdot \sqrt[4]{g\delta\rho\rho^*} \quad (6)$$

where α_k and α_r are the convective and radiative heat transfer coefficients, V_0 and V – volume of a spheroidal drop at initial time and time t, δ – thickness of an oblate spheroid representing the droplet (D – its diameter), T_{sat}^* - vapor saturation temperature, T_{wall} – wall temperature, $\langle T\rangle$ – average temperature of the droplet, ρ and ρ^* are magnesium liquid and vapor densities, λ^* - vapor thermal conductivity, C_p – liquid magnesium heat capacity, r is the latent heat of vaporization (liquid-vapor phase change enthalpy), $g = 9.81\ \mathrm{m^2s^{-1}}$. Figure 2, right shows the normalized dynamics of the magnesium jet velocity and the buildup of reactor pressure during injection through a nozzle with a diameter $\sigma_N = 0.2$ mm into the reactor with internal pressure of 50 Pa from a feed tank with a hydraulic head $h = 0.4$ m and overpressure $P = 0.2$ bar. Here, the pressure of magnesium vapor with density m_{Mg}/10 L is used as the characteristic pressure and the steady-state outflow velocity at constant pressure in the reactor - as the characteristic velocity. The dispensing time τ_p required for supplying the predetermined amount m_{Mg} of magnesium into the reactor is obtainable from eqn (5) and the relation

$$m_{Mg} = 0.25\,\pi\sigma_N^2\rho\int_0^{\tau_p} v(t)dt \quad (7)$$

For the given parameters τ_p is 2.3 sec.

2.2. Injection of titanium tetrachloride and reaction

In the next stage a corresponding stoichiometric amount of titanium tetrachloride $TiCl_4$ is injected through a narrow aperture into the reaction retort containing vaporized magnesium. Flowing through the feed line to the lid of the reactor the $TiCl_4$ will be heated by the channel walls. The estimations taking into account a possible thermal insulation yield a temperature in excess of 400 K for a $TiCl_4$ jet entering the chamber with magnesium vapours. In this case, the saturated vapour pressure of $TiCl_4$ is much greater than that of magnesium (Figure 3, left) resulting in rapid vaporization. During the expansion the vapour-liquid jet of $TiCl_4$ reacts with the magnesium vapours, heats up and disintegrates into droplets of liquid magnesium chloride containing solid titanium particles and $TiCl_2$. The reaction products are then deposited at the walls and bottom of the retort.

2.3. Removal of vapors of magnesium chloride from the reactor

Upon the completion of the reaction, all of the magnesium dichloride $MgCl_2$ and excess magnesium will remain in the gas phase in the reactor. As soon as the valve connecting the reactor to the condenser is opened, the vapors from the reactor are sucked into the condenser. The hydraulic resistance of this channel is the main obstacle for the vapor removal rate. The pressure drop in the channel is related to the mass transfer rate as

$$p_V(t) - p_{cond} = \left(\lambda\frac{L}{D_H} + \sum\zeta_i\right)\cdot\frac{\rho(t)v(t)^2}{2} \quad (8)$$

where $p_V(t), \rho(t)$ are the vapor pressure and density in the reactor at a time moment t, p_{cond} is the saturated vapour pressure of $MgCl_2$ in the condenser (at 720 °C), L and D_H are the length and the hydraulic diameter of the channel, ζ_i denotes local hydraulic resistances [15], $v(t)$ is the vapor velocity in channel. With the diameter of the channel being 5 cm, length 2 m and with the cross-section narrowing at the valve to 2.5 cm, the estimated time of vapor evacuation from the reactor into the condenser will be about 0.095 s.

2.4. Scenario of the reactor operation

Figure 3, right shows a tentative flow chart for a single cycle. The cycle duration includes the time of evaporation of the supplied magnesium, the time of the reaction, the precipitation of fine dispersed titanium on the walls, and the time of the distillation of volatiles into the condenser. The titanium is periodically spilled through the bottom hole by opening the shutter. The quantity of titanium per cycle per 10 L of reactor volume is equal to a cube with the edge 0.47 cm. Powder titanium can undergo electroslag remelting to produce ingots of titanium alloys.

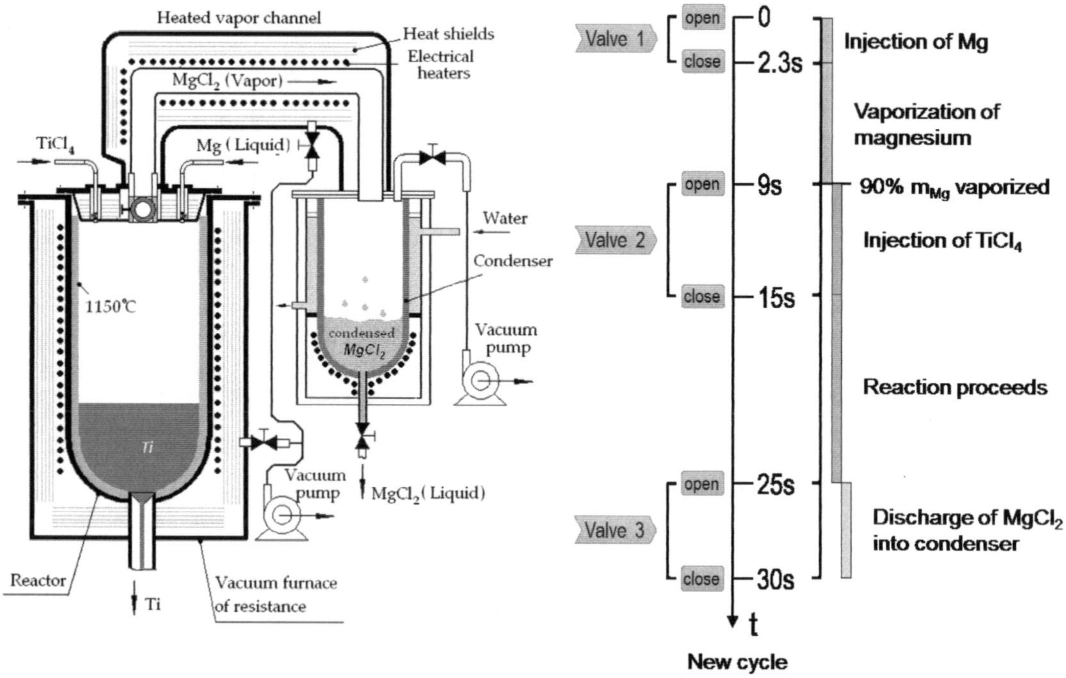

Figure 3. Left - basic components of the proposed device for intensive production of titanium by cyclic magnesiothermic reduction from titanium tetrachloride; right – tentative flow chart of one operating cycle: **(initial state)** – vacuum, reactor temperature 1150 °C; **(step 1)** – injection of magnesium, 0.333g per 10 L of reactor volume, the pressure in the reactor increases to 0.163 bar; **(step 2)** – injection of 1.297 g $TiCl_4$ (per 10 L), the reaction proceeds rapidly with heat release, 1.302 g (per 10 L) of $MgCl_2$ vapors and 0.327 g (per 10 L) of solid titanium are formed in the reactor, the pressure is 0.137 bar; **(step 3)** – the outlet valve is opened, which connects the reactor to a cold evacuated chamber (condenser) with the temperature 720 °C (the vapor pressure is 48 Pa) to discharge $MgCl_2$; **(end of cycle)** – the cycle is completed after the valve is shut (closed), the pressure in the reactor is 50 Pa and the cycle can be repeated.

3. Conclusions

We attempted to predict how to utilize the technological advantage of the reaction occurring in the reactor at 1150 °C predominantly in the gas phase and how to realize the process using special equipment, i.e. high-temperature, quick-operating valves and gates or shutters. This implies a successive batch feeding of the reagents and the removal of volatiles after the reaction, making room for another reaction. In this case, the working space will be reduced only due to the titanium remaining in the reactor. Since in the gas phase titanium is formed as powder [12], then, apparently, it will be possible to design a reactor with the periodic titanium powder spilling through the opening channel into a separate vessel outside the reactor and the furnace. Then the thermal reduction of titanium by magnesium can be assumed continuous. Currently a prototype of the reactor with operating temperature up to approx. 1200 °C, combined with the condensing unit, are being built to carry out experiments on the reduction of titanium tetrachloride with dosed supply of reagents and periodic distillation. Preliminary estimates indicate that this process could be economically viable in comparison with the current industrial practice, provided that a number of technical difficulties could be resolved. The design of the structural units supplying the reagents through the lid of the reactor including high-temperature fast-switching valves and protection of the nozzles from corrosive environment in the reactor chamber will be key.

References

[1] van Vuuren D S 2015 Direct titanium powder production by metallothermic processes *Titanium Powder Metallurgy* (Elsevier) 69–93
[2] Wartman F S, Baker D H, Nettle J R and Homme V E 1954 Some Observations on the Kroll Process for Titanium J. Electrochem. Soc. 101(10) 507-513
[3] Nagesh Ch R V S, Sridhar Rao Ch, Ballal N B, Krishna Rao P 2004 Mechanism of titanium sponge formation in the kroll reduction reactor *Metallurgical and Materials Transactions B* 35(1) 65-74
[4] Garmata V A et al. 1983 *Titanium* (Moscow: Metallurgiya) 559
[5] Olkhov Yu N, Ogurtsov S V et al. 1969 Thermodynamics of the titanium redox reaction from titanium tetrachloride by magnesium and the choice of optimal ways to intensify the titanium production *Investigations in Titanium Chloric Metallurgy* (Moscow: Metallurgiya) 302
[6] Ivanov S L, Foliforov V M, Platacis E et al. 1994 Special features of Electromagnetic Pump Investigations for Space Nuclear Reactors *Proceedings of 2nd International Conference of Energy Transfer in MHD-flow*. Aussois-France, 643-650
[7] Byalobzheskiy A V, Tsirlin M S, Krasilov B I 1977 *High-temperature corrosion and the protection of super-refractory metals* (Moscow: Atomizdat)
[8] Ivanov S L, Maiorov M M, Blumberg E, Zyk A A reactor for the accelerated magnesium exothermic reduction of titanium from titanium tetrachloride (in preparation)
[9] Leland J D 1996 Economically producing reactive metals by aerosol reduction JOM 48(10), 52-55
[10] Hansen D A, Gerdemann S J 1998 Producing titanium powder by continuous vapor-phase reduction JOM 50(11), 56-58
[11] Sohn H Y 1998 Ti and TiAl powders by the flash reduction of chloride vapors JOM 50(9), 50-51
[12] Turner P C, Hartman A, Hansen J S, Gerdemann S J Low cost titanium - myth or reality. DOE/ Albany Research Center - 2001-086
[13] Lide D R 2003 *CRC Handbook of Chemistry and Physics*, 84th Edition (CRC Press)
[14] Tarasov A V 2003 *Titanium Metallurgy* (Moscow: Academkniga) p 328
[15] Idelchik I E 2012 *Handbook of hydraulic resistance* (U.S. Atomic Energy Commission, translated from Russian)
[16] Levich V G 1962 *Physicochemical hydrodynamics* (Englewood Cliffs, N.J.: Prentice-Hall)
[17] Lielausis O, Klyukin A A, Platacis E, Peinbergs J 2015 MHD experiments on liquid metal jet-like flows guided by curved substrates *Magnetohydrodynamics* 51(4) 685-694
[18] Kutateladze S S 1979 *Fundamentals of the theory of heat exchange* (Moscow: Atomizdat)
[19] Zhang W, Zhu Z, Cheng C Y 2011 A literature review of titanium metallurgical processes *Hydrometallurgy* 108 (2011) 177-188

Acknowledgments
This work was supported by the ERDF project No. 1.1.1.1/16/A/085 (Ti-REMOLD).

IOP Conf. Series: Materials Science and Engineering **355** (2018) 012011 doi:10.1088/1757-899X/355/1/012011

Direct chill casting of aluminium alloys under electromagnetic interaction by permanent magnet assembly

Andris Bojarevičs, Imants Kaldre, Mikus Milgrāvis, Toms Beinerts

Institute of Physics University of Latvia, Miera 32, LV-2169, Salaspils, Latvia

Imants.Kaldre@lu.lv

Abstract. Direct chill casting is one of the methods used in industry to obtain good microstructure and properties of aluminium alloys. Nevertheless, for some alloys grain structure is not optimal. In this study, we offer the use of electromagnetic interaction to modify melt convection near the solidification interface. Solidification under various electromagnetic interactions has been widely studied, but usually at low solidification velocity and high thermal gradient. This type of interaction may succeed fragmentation of dendrite arms and transport of solidification nuclei thus leading to improved material structure and properties. Realization of experimental small-scale crystallizer and electromagnetic system has been described in this article.

1. Introduction

Aluminium alloys microstructure and physical properties may vary significantly depending on solidification velocity and temperature gradient at the solidification interface, and melt flow near solidification interface as well. In industry, aluminium alloys for extrusion and forging is prepared by using crystallizers, which ensure controlled solidification velocity and oriented solidification front direction [1]. There are several technical solutions how to achieve rapid solidification up to 10 mm/s and large diameter ingots. One of the most prevalent is direct chill casting, where heat is evacuated by spraying oil and water at the solid part of the ingot [2]. It has been demonstrated that pulsed or AC electromagnetic interaction during directional solidification can significantly modify solidification structure of aluminium alloys [3]. Electromagnetic force may influence the melt flow in several ways. Static magnetic field has two effects. Firstly, it damps the melt flow perpendicular to magnetic field lines, and, secondly, magnetic field interacts with electric current near the solidification interface and drives melt convection [4,5] and can deform the solidified dendrite arms [6]. Electric current may arise because of thermoelectric effect between solid and liquid phases in presence of temperature gradient along solidification interface [7] or current can be directly the melt, or induced by the alternating magnetic field. Diverse types of melt flow give different effects on solidification structure. For example, large scale flow may change solidification interface shape and influence the structure difference between core and outer crust of the ingot [8]. Whereas small dendrite-scale flow may succeed composition homogeneity and alter dendrite morphology. However, there is no direct link between melt convection during solidification and solidification microstructure, thus for each alloy and electromagnetic interaction series of experiments should be done.

Content from this work may be used under the terms of the Creative Commons Attribution 3.0 licence. Any further distribution of this work must maintain attribution to the author(s) and the title of the work, journal citation and DOI.
Published under licence by IOP Publishing Ltd

2. Experimental setup

Small scale direct chill casting experimental setup has been developed in MHD technology laboratory of Institute of Physics University of Latvia. Principal scheme of experimental setup is shown in Figure 1. Aim of the experimental setup is to test the possibility to influence solidification structure and impurity distribution in aluminium ingot by various electromagnetic interactions. Vertical DC magnetic field is applied by the permanent magnet system placed around the solidification zone. Solidification interface takes place in graphite tube at the middle of permanent magnet system where field induction reaches maximal value. Just below graphite tube the solid part of the aluminium rod is cooled by water spray. DC magnetic field interacts with injected electric current in the aluminium near the solidification interface.

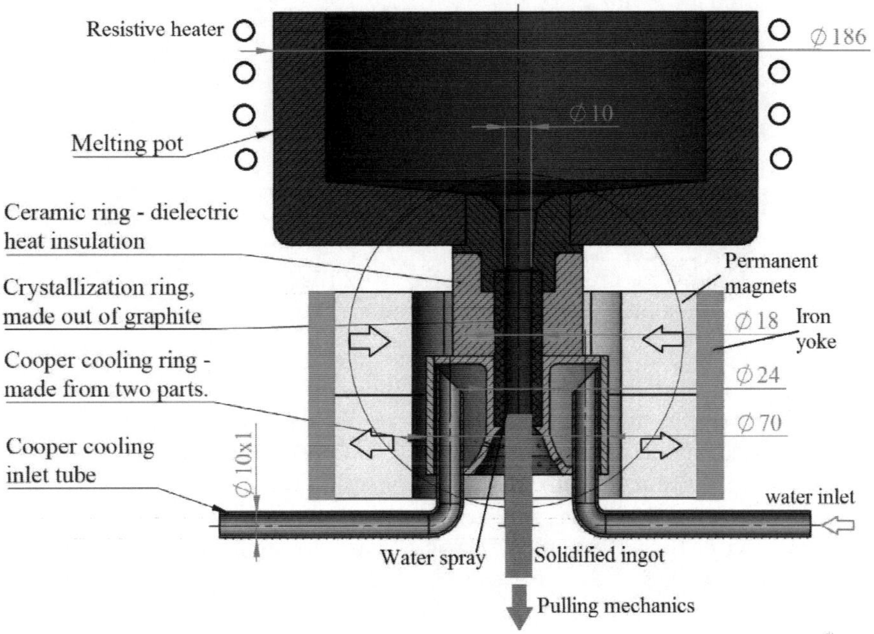

Figure 1. Principal scheme of the experimental setup.

3. Results and discussion

For the first experiments, it is planned to apply axial electric current through the solidification interface. In this case in the bulk of the solid and liquid parts magnetic field and current are parallel, thus there is no Lorentz force. However, at the dendritic solidification interface current redistribution takes place due to significant difference in electric conductivities between solid and liquid phases. It has been shown by previous work that such type of interaction may induce liquid phase flow at dendrite scale and modify solidification microstructure and segregation of the components. Parameters of experimental system has been chosen to induce electromagnetic forces in the melt comparable to similar studies described in literature [9].

In this work planned current density is starting from 1 A/mm^2. Electric current at the dendritic solidification interface is redistributed as shown in Figure 2. In this case current component perpendicular to magnetic field creates Lorentz force and local convection vortices around each dendrite arm is created. Typical primary dendrite scale of aluminium is 100 μm. Melt convection velocity can be estimated by balancing electromagnetic and viscous forces in simplified Navier-Stokes equation [9]. For aluminium 0.5 T axial magnetic field would give characteristic velocity of few centimetres per second, which could have significant effect at solidification velocities 1-3 mm/s.

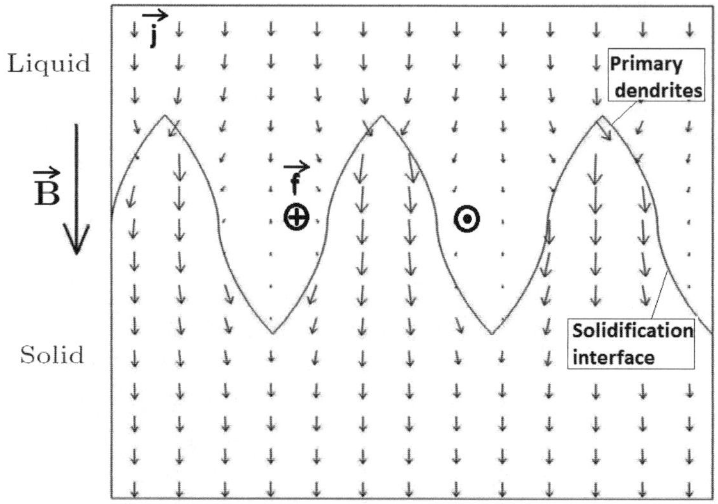

Figure 2. Electric current distribution at the dendritic solidification interface (B-magnetic field induction, j-current density, f-force density).

Permanent magnet system aimed to achieve 0.5 T is assembled from segment magnets. Iron yoke to optimize magnetic flux as shown in Figure 3(a). Inner cavity of magnet system is 80 mm in diameter and axial magnetic field induction at the middle is about 0.45 T as shown in Figure 3(b) by comparing measurements and numerical model. Low carbon soft iron yoke guides magnetic flux with minimal resistance thus magnetic field induction in the magnet is maximized. For these N42M type magnets maximum work temperature is 100 °C, thus magnets are protected from excess heat from hot zone by copper screen which is water cooled. Thus, magnets can be easily placed within several centimetres from molten aluminium. Permanent magnet systems can be designed for many applications, such as liquid metal pumps and stirrers, couplings and bearings, sensors, motors etc. [10,11] Halbach array principle is used in systems to increase and improve magnetic field distribution by creating permanent magnet arrays from many smaller magnets. Ways to optimize magnetic flux has been described by several authors [12, 13].

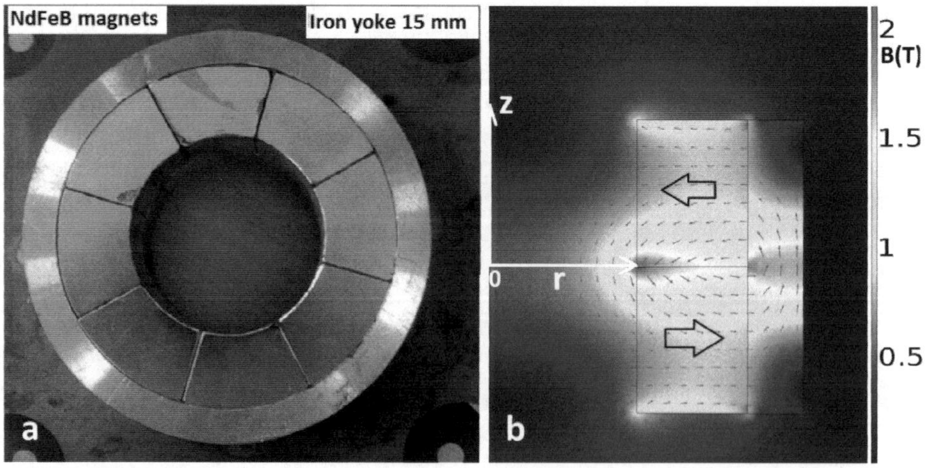

Figure 3. Magnet system: a) Assembled permanent magnet system; b) axysymmetric Comsol model (Magnetic field induction) Black arrows indicate magnetization direction.

IOP Conf. Series: Materials Science and Engineering **355** (2018) 012011 doi:10.1088/1757-899X/355/1/012011

Comparison between Comsol 5.0 numerical model and gaussmeter measurements shown in Figure 4. demonstrates that numerical calculations are accurate. Axysymmetric stationary model is solved to calculate the magnetic flux density. Yoke is assumed to be soft iron with saturation magnetization of 2.1 T, remanent flux density of the NdFeB magnets is 1.4 T. Triangular mesh consisting of 46000 elements is used for calculations. Difference may be explained by the facts that there are small gaps between magnet pieces and that magnetization is not entirely homogeneous. In our case ingot diameter is 10 mm thus inhomogeneity of magnetic field does not play a crucial role. If such magnetic field is provided by electromagnet, then number of ampere-turns necessary can be estimated from finite coil magnetic field induction formula:

$$B_z(z) = \frac{\mu_0 NI}{\sqrt{h^2 + R^2}}$$

Where h is height of the magnet, R is coil radius, μ_0 is vacuum permeability. For our case (h=10cm, R=4 cm, R_{outer}=9 cm) we get necessary ampere-turns IN=30 kA turns. Such electromagnet would be difficult to make and operate thus in this case permanent magnet assembly is definitively an advantage.

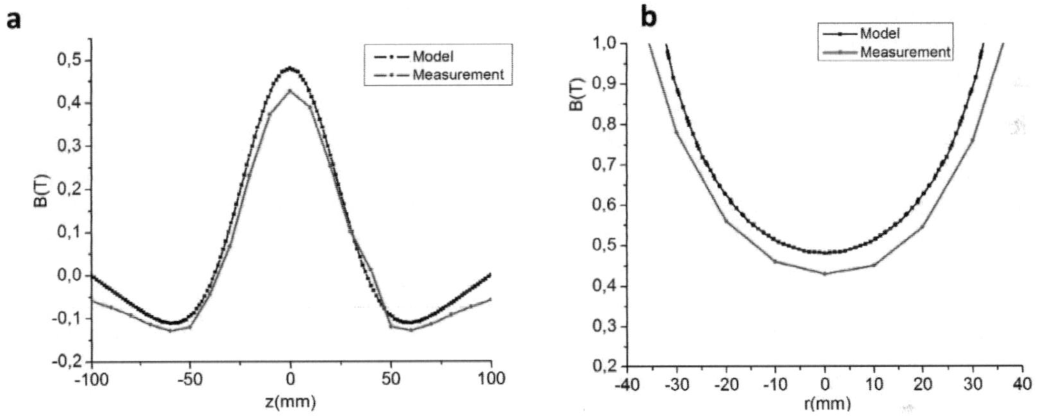

Figure 4. Comparison of numerical model and measurements. Magnetic field induction: a) Along axis of the magnet system; b) along radial position at the middle height.

NdFeB magnets with remanent flux density of 1.4 T are becoming more available nowadays, even in custom shapes. Limitation is size in magnetization direction which is around 5 cm. Larger magnet system with 20 cm diameter can be built in the same manner. Such system would require 8 liters of NdFeB magnets and 2 cm thick iron yoke to prevent oversaturation. Estimated price of the magnets would be around 3000 Eur. Total weight of the magnet system with yoke would be around 80 kg. Magnet system sizes are shown on Figure 5(a). Calculated flux density along z axis and along diameter are shown in Figure 5(b). More than 0.4 T can be achieved in 20 cm diameter magnet. With air filled electric coil this would require 71 kA turns. Thus, even for a larger application permanent magnet assembly is reasonable alternative for complicated electromagnet.

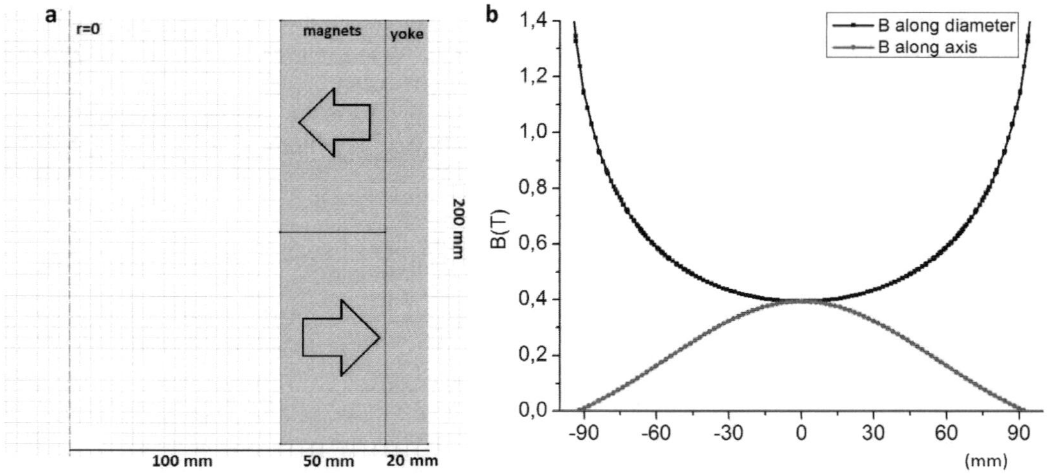

Figure 5. Magnet system for 200 mm diameter axial field assembled form segment magnets and iron yoke: a) cross section of the magnet system (with magnetization directions); b) magnetic field induction.

4. Conclusions

Direct chill casting is one of the promising technologies for high quality aluminium alloy production maintaining certain properties. Our current work is aimed to improve existing technology by investigating the potential applications of electromagnetic interaction to the process. Electromagnetic interaction on liquid aluminium is perspective because of relatively low melting temperature, high electrical conductivity and low density of the material. Thus, electromagnetic elements can be placed close to the liquid metal and smaller magnetic field amplitude is necessary to achieve sufficient force density in the melt. Designed permanent magnet assembly for DC field has shown its potential and in the future permanent magnets could replace superconducting magnets in several applications where moderate field is necessary.

Acknowledgements

This work is supported by University of Latvia effective cooperation project: "Use of combined electromagnetic method for continuous casting for improved light metal alloys".
ERDF project „Mechanical engineering competence centre", ID Nr.1.2.1.1/16/A/003, "Development of electromagnetic crystallizer for special fine-grained alloy casting"

References

[1] Grandfield J.F, McGlade P.T 1996 *DC Casting of Aluminium: Process Behaviour and Technology.* Comalco Research Centre, 15 Edgars Rd. Victoria 3074, Australia

[2] Schiffl A 2015 *Direct Chill Casting of Magnesium using Hot-Top-Technology for Improved Casting Quality.* ARC Leichtmetall kompetenz zentrum Ranshofen GmbH (LKR), Ranshofen

[3] Wang R.Z, Qi J.B, Wang B, Zhang W, Wang J.Z 2016 *Journal of Materials Processing Technology* Solidification behavior and crystal growth mechanism of aluminium under electric pulse. **237**, pp 235-242

[4] Vives Ch, Perry C 1987 *International Journal of Heat and Mass Transfer* Effects of magnetically damped convection during the controlled solidification of metals and alloys, **30**, Issue 3, pp 479-496

[5] Wang J, Fautrelle Y, Ren Z.M, Li X, Nguyen-Thi H, Mangelinck-Noel N, Salloum Abou Jaoude G, Zhong Y.B, Kaldre I, Bojarevics A, Buligins L 2012 *Applied Physics Letters* Thermoelectric magnetic force acting on the solid during directional solidification under a static magnetic field, **101** (25), no. 251904

[6] Jianbo Yu, Dafan Du, Zhongming R, Fautrelle Y, Moreau R, Li X 2017 *ISIJ International* Influence of an Axial Magnetic Field on Microstructures and Alignment in Directionally Solidified Ni-based Superalloy, **57**, No. 2, pp 337-342

[7] Kaldre I, Fautrelle Y, Etay J, Bojarevics A, Buligins L 2010 *Magnetohydrodynamics* Investigation of liquid phase motion generated by the thermoelectric current and magnetic field interaction **4**, pp 371-380.

[8] Wang J, Fautrelle Y, Nguyen-Thi H, Reinhart G, Ren Z 2016 *Metallurgical and Materials Transactions A* Thermoelectric Magnetohydrodynamic Flows and Their Induced Change of Solid–Liquid Interface Shape in Static Magnetic Field-Assisted Directional Solidification **47**, Issue 3, pp 1169–1179

[9] Kaldre I, Bojarevics A, Fautrelle Y, Etay J, Buligins L 2012 *Magnetohydrodynamics* Current and magnetic field interaction influence on liquid phase convection **48 (2),** pp 399-406.

[10] Coey J.M 2002 *Journal of Magnetism and Magnetic Materials* Permanent magnet applications **248**, Issue 3, pp 441-456

[11] A Bojarevičs, R Baranovskis, I Kaldre, M Milgrāvis and T Beinerts 2017 *IOP Conf. Ser Mater. Sci. Eng* Two cylinder permanent magnet stirrer for liquid metals **228** 012022

[12] K Halbach 1980 *Nuclear Instruments and Methods* Design of permanent multipole magnets with oriented rare earth cobalt material **169**, Issue 1, pp 1-10

[13] Zhu Z.Q, Howe D 2001 *Electric Power Applications, IEE Proceedings* Halbach permanent magnet machines and applications **148** pp 299 - 308

IOP Conf. Series: Materials Science and Engineering **355** (2018) 012012 doi:10.1088/1757-899X/355/1/012012

Numerical Simulation of Induction Channel Furnace to Investigate Efficiency for low Frequencies

N Tran Thi Hang and U Lüdtke

Faculty of Electrical Eng. and Information Tech., Technische Universität Ilmenau
Gustav-Kirchhoff-Str. 1, D-98693 Ilmenau, Germany

E-mail: ninh.tran-thi-hang@tu-ilmenau.de

Abstract. The foundry industry worldwide commonly uses induction channel furnaces to heat and melt alloys. The operating frequency is one of the main issues when constructing an efficient channel furnace. It is possible to choose operating frequencies lower than 50 Hz using a modern IGBT power converter. This work shows the simulation results using ANSYS with the goal of finding the best electrical frequency necessary to operate the induction furnace. First, a two-dimensional model is used to calculate the efficiency depending on frequency. Then, the channel model is extended to a more realistic three-dimensional model. Finally, the influence of frequency, inductor profile, and several components of the induction channel furnace are discussed.

1. Introduction

The basic structure of an induction channel furnace is equal to a transformer (Fig. 1). The short-circuited secondary coil with one winding is the melting channel where Joule heat is generated and used to heat the melt in the tank above. The excellent coil flux guide for the magnetic field results in a very good flux coupling between the inductor coil and the melting channel which represents the secondary coil. This is an important reason for the good electrical efficiency of the induction channel furnace in comparison to the induction crucible furnace. There are further problems to consider when operating an induction channel furnace. We aim to prevent undesirable erosions or depositions at the surface of the refractory material. The heat is generated in the channel and must be transported to the melt in the tank. The power input to the channel is limited to prevent overheating.

A main goal when operating induction channel furnaces is to improve life time and energy savings in order to reduce costs and to increase competitiveness. For many years researchers have continually optimized the components of induction channel furnaces, aiming to optimize melt flow and Lorentz forces, to reduce heat losses in the housing or to construct an optimal melting channel shape. Using frequencies less than 50 Hz is an interesting possibility for improving the electrical efficiency, since modern low-cost generators with Insulated-Gate Bipolar Transistors are available [2-5]. There is no induction channel furnace available to carry out experimental investigations since a great disadvantage is the fixed geometry. It would take a great effort to modify the channel shape and the inductor profile. The frequency range is determined by the selected generator. Thus, numerical simulations using finite elements are very suitable to investigate the influence of low frequencies on the electrical efficiency.

Content from this work may be used under the terms of the Creative Commons Attribution 3.0 licence. Any further distribution of this work must maintain attribution to the author(s) and the title of the work, journal citation and DOI.
Published under licence by IOP Publishing Ltd

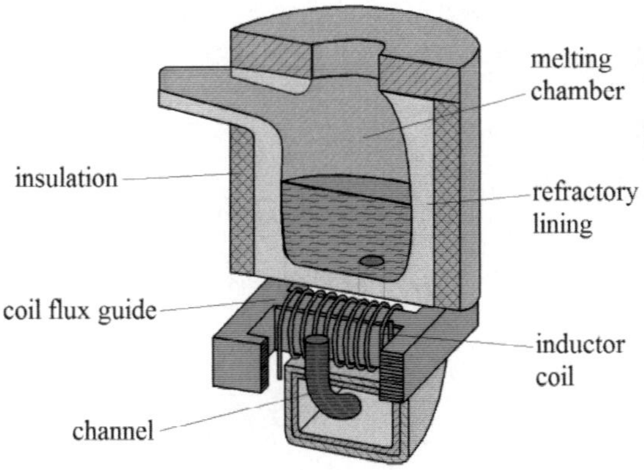

Figure 1. Simplified sketch of an induction channel furnace [1].

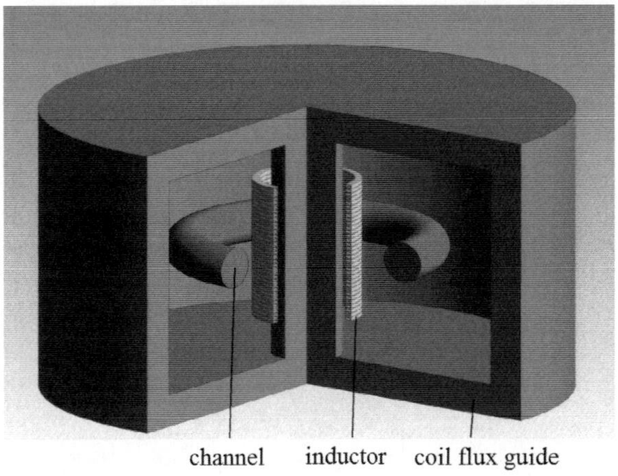

Figure 2. The simplified axisymmetric shell-core model.

2. The Axisymmetric Shell-Core Model

We aim to investigate the interplay between channel shape and inductor coil depending on frequency. That's why the inductor coil profile is considered in detail. Since a three-dimensional inductor coil model is very complex and time-consuming, we use the simplified axisymmetric shell-core model (see Fig. 2). The axisymmetric shell-core works as coil flux guide to ensure the excellent interlink between inductor coil and channel. The melting tank and the pass from the channel to the tank do not exist for these calculations. It is not possible to consider hysteresis loss and eddy current loss in the shell-core in the calculations. Further, the housing and the cooling shell geometry cannot easily be transformed into the axisymmetric model. Thus, eddy current loss is neglected as well. Numerical variables such as the winding pitch, inductor inlet, and housing elements must also be neglected. This axisymmetric model represents a cross-section of the magnetic flux guide, the channel, and the inductor, which are 360° around the Z-axis. This model aims to reduce the computing effort while maintaining sufficient numerical calculation accuracy. Therefore, for our simulations we use the one-loop channel furnace model, which is the simplest induction channel furnace.

The Maxwell equations are solved using the so-called magnetic vector potential and the Finite Element Method (FEM). To simulate the shell-core model, the in-house code PROMETHEUS is used [4]. Additional calculations are done using the commercial software-program ANSYS-MAXWELL [7]. The FEM-theory is state-of-the-art and a large number of publications are available regarding this topic. Detailed investigations regarding the shell-core model are published in [8].

3. Simulation of the Induction Channel Furnace

Experimental investigations using low inductor current frequencies below 50 Hertz for induction channel furnaces have shown that it is possible to improve the electrical efficiency [6]. This improved electrical efficiency could be caused by the hysteresis loss in the coil flux guide and eddy current losses in the housing, which are both reduced when the frequency decreases. The influence of the frequency on the current density distribution in the inductor and in the melt depends on geometry and shall be investigated for a special kind of induction channel furnace. For this work, the construction data of a 250 kW channel furnace with one loop for melting cast iron was considered. This type of channel furnace was investigated in the famous BMFT - joint research project [2].

3.1. The Axisymmetric Shell-Core Model

Using the geometry data, the one-loop induction channel furnace will be transformed into the simplified axisymmetric shell-core model (see Fig. 2). To simulate, we use a very fine mesh, which results in a very highly accurate electrical efficiency calculation depending on frequency.

IOP Conf. Series: Materials Science and Engineering **355** (2018) 012012 doi:10.1088/1757-899X/355/1/012012

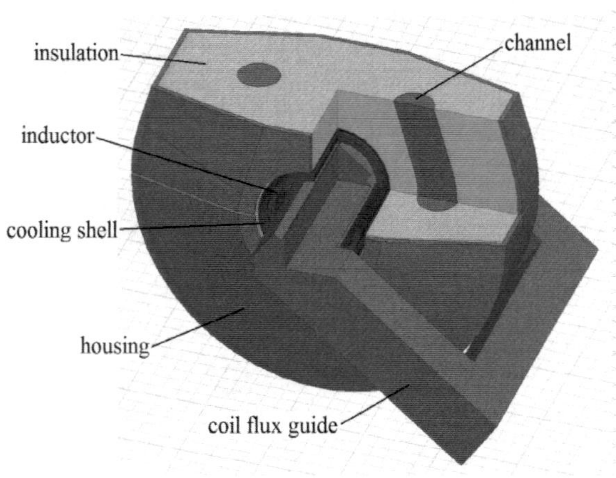

Figure 3. The three-dimensional model with housing and cooling shell (sketch without melting tank).

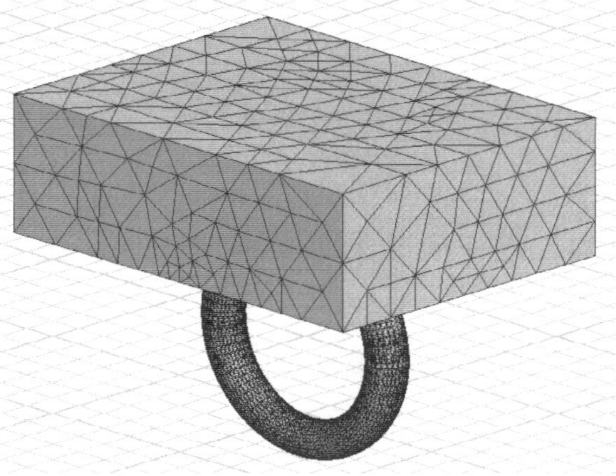

Figure 4. Mesh of simplified tank and channel (MAXWELL).

3.2. The Three-Dimensional Model

Our goal is to predict the influence of the simplifications coming from the shell-core model and to increase the model accuracy. To achieve this goal, we use the program ANSYS-MAXWELL to carry out a three-dimensional simulation of the induction channel furnace including housing, cooling shell, and realistic flux guide for the magnetic field (see Fig. 3). Fig. 4 shows the three-dimensional adapted mesh of the melting channel and the melting tank. The mesh is typical for MAXWELL calculations. In this case the element size is partially much coarser than in the case of the two-dimensional axisymmetric simulation of the shell-core model.

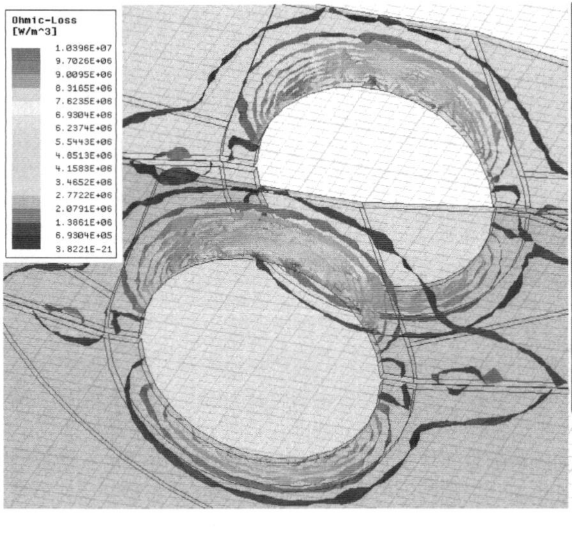

Figure 5. Joule heat in housing.

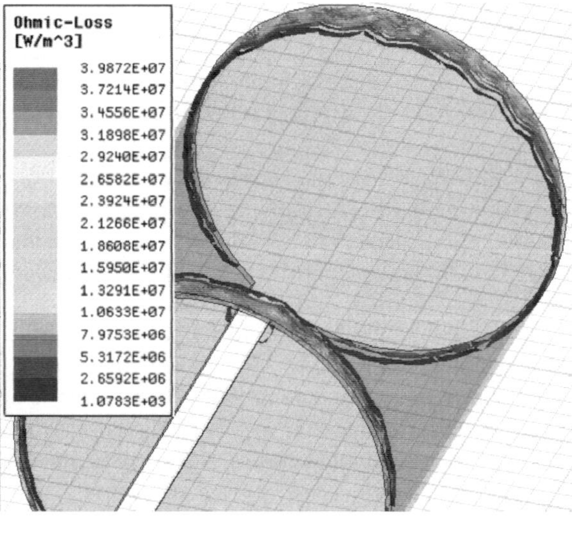

Figure 6. Joule heat in cooling shell.

4. Simulation Results

Detailed results of the two-dimensional shell-core model investigations are published in [8]. One important improvement of the three-dimensional model is considering the Joule heat in the housing (see Fig. 5) and in the cooling shell (see Fig. 6). Usually FEM-programs (MAXWELL, PROMETHEUS) are able to parameterize particular quantities. Mesh, material properties, and impressed currents are assumed to be constant and finally the frequency is the parameter.

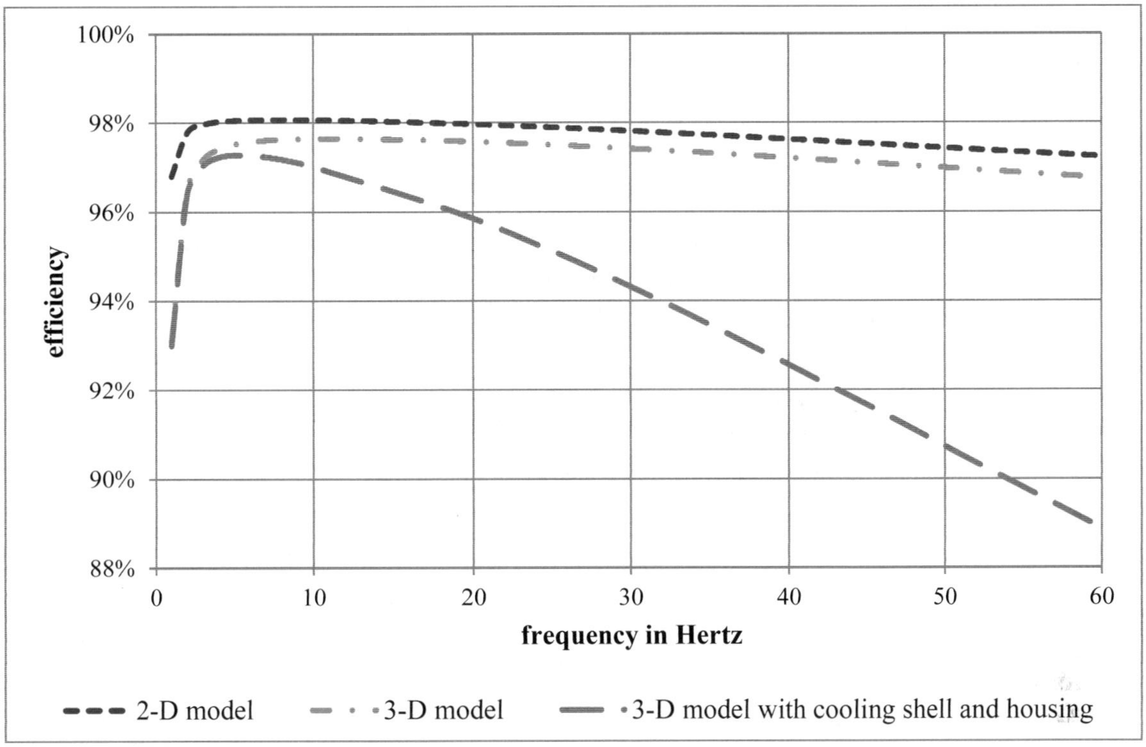

Figure 7. Comparison of the two-dimensional shell-core model and three-dimensional model.

The frequency is divided in 1 Hz steps beginning at 1 Hz up to 60 Hz. The simulation will be done for every frequency step and all important quantities will be saved. The frequency-dependent electrical efficiency is shown in Fig. 7. Regarding the shell-core model (two-dimensional axisymmetric) we find the highest electrical efficiency of 98 % at the remarkably low frequency of 9 Hz. We further aim to evaluate the accuracy of the shell-core model using a three-dimensional simulation without housing and without cooling shell since the shell-core model is not able to consider these slotted structural elements. Even though the three-dimensional model uses a partially much coarser mesh we can observe a relatively good agreement of the electrical efficiency depending on frequency (see Fig. 7: 2-D model – 3-D model). The maximum electrical efficiency of the three-dimensional simulation without housing and without cooling shell is equal to 12 Hz. In the frequency range lower than 10 Hz the difference between both curves increases. In the frequency range higher than 10 Hz the distance between both curves is nearly constant. The difference can be explained by the total resistance of the channel. The total electrical resistance of the channel combined with the melting tank (three-dimensional model) is lower than the total electrical resistance of the axisymmetric shell-core model. The lower resistance leads to a smaller electrical efficiency.

The third curve of the three-dimensional model in Fig. 7 shows the electrical efficiency depending on frequency of the complete model including housing and cooling shell. The maximum electrical efficiency can be seen at a surprisingly low value of 5 Hz. Upon increasing the frequency, the electrical efficiency decreases stronger than the electrical efficiency of the shell-core model. The only reason can be the Joule heat in the housing and cooling of the three-dimensional model, which increases depending on frequency.

Two -dimensional simulations are better suited for our purpose, as can be seen by comparing the calculations. The 2-D model (MAXWELL, 340MB memory, about 53 000 elements) needs about 1 minute to reach an error lower than 0.01 %. The 3-D model (MAXWELL, 8.35 GB memory, about 1 370 000 elements) needs about 3.5 hours to fall below an error of 1 %. In this case it is not possible to achieve the error limit of 0.01 %. The calculations were done using a modern PC with an Intel® Xeon® Processor E5-2650 (2.3 GHz) and 758 GB memory.

5. Conclusions

This work aims to provide the numerical simulation of the frequency dependence with regard to the electrical efficiency of induction furnaces. Using the axisymmetric shell-core model, numerical simulations are well suited to investigate the influence of the inductor profile, of different channel cross-sections or sizes, and of different element positions. All elements of the model can be described using a very fine mesh in order to achieve highly accurate simulation results. The advantage of the two-dimensional simulation is the acceptable calculation time needed to compute the given frequency range step by step, ensuring a low simulation error. The systematical error, caused by the simplifications of the two-dimensional shell-core model must be accepted. Three-dimensional simulations are done to verify the shell core-model. The comparison shows a good agreement between the electrical efficiency depending on frequency. Differences can be explained by the simplifications of the shell-core model. The maximum efficiency lies at nearly the same position.

The copper coil profile can be modeled very exactly using the 2-D model. Using the 3-D model, the coil profile can only be approximated with a coarse mesh. The 2-D model and 3-D model results are in good agreement. As a consequence, the 2-D model is ideal for studying different types of special copper profiles.

The realistic three-dimensional model including the housing and the cooling shell provides further arguments for the use of low frequencies to improve the electrical efficiency of induction channel furnaces. Further work is needed to consider thermal field effects, since the overall efficiency of the induction heating system is a combination of both thermal and electrical efficiency. The disadvantage of the three-dimensional model is the high computation time to calculate the electrical efficiency for all frequencies.

The induced power is an important condition which we have to consider. It is basically possible to use a frequency of 10 Hz. Using a frequency of 5 Hz is critical because the induced power starts to decrease rapidly. But there is another non-technical problem. The channel furnace produces noise and a frequency below 25 Hz generates painful infrasound, so that manufacturers won't apply frequencies below 25 Hz. If future casting houses are fully automated, then applying frequencies below 25 Hz will greatly increase electrical efficiency.

References

[1] 1937 *Russ-Elektroofen K.G. Laboratory Report:* Niederfrequenzinduktionsschmelzöfen für Schwermetalle, Leichtmetalle und Eisen (Köln: Russ-Elektroofen K.G.) vol 6 chapter 2

[2] Andree W, Drewek R, Eggers A, Lüdtke U, Nacke B and Walther A 1993 *Verbesserung der Energieausnutzung in umweltfreundlichen Induktions-Rinnenöfen durch Optimierung der Schmelzenströmung* (BMFT-Research joint project University of Hannover, ABB Dortmund, Technical University of Ilmenau) Project-Nr. 0328777A-C

[3] Nacke B, Idziok K-H and Walther A 1999 *Proc. Int. Conf. on Modelling of Material Processing* New development of High-Powered Channel Inductors for Melting Brass (Riga) pp 200-209

[4] Weigel W 2002 *Zur numerischen Berechnung von Wirbelstromverlusten in Konstruktionsteilen von Induktionsöfen* Dissertation, Technische Universität Ilmenau, Department of Electrical Engineering and Information Technology

[5] Drysch P, Niemann M and Pfaffenhöfer U 1997 Neues Induktorkonzept für Rinnenöfen *Gießerei, Die Zeitschrift der Deutschen Gießereivereinigungen* vol 1 pp 1-9

[6] Schuller R 2012 Application of induction heating in Low-Frequency (*Proc. on Workshop Elektroprozesstechnik*) Technische Universität Ilmenau, Förderkreis Elektrowärme Ilmenau e.V.

[7] 2016 http://www.ansys.com

[8] Lüdtke U and Tran Thi Hang N 2014 Numerische Simulation der Frequenzabhängigkeit des Wirkungsgrades bei Induktionsrinnenöfen im Niederfrequenzbereich *ewi - elektrowärme international* vol 4 pp 65-68

IOP Conf. Series: Materials Science and Engineering **355** (2018) 012013 doi:10.1088/1757-899X/355/1/012013

Comparison of Electromagnetic and Marangoni Forces on Thin Coatings during Rapid Heating Process

T Steinberg[1], T Opitz[2], A Rybakov[1], E Baake[1]

[1] Institute of Electrotechnology, Leibniz Universität Hannover, D-30167 Hanover, Germany
[2] Volkswagen AG, D-38440 Wolfsburg, Germany

steinberg@etp.uni-hannover.de

Abstract. The present paper is dedicated to the investigation of Marangoni and Lorentz forces in a rapid heating process. During the melting of aluminum-silicon (AlSi) layer on the bor-manganese steel 22MnB5, the liquid AlSi is shifting from the middle to the side and leaves dry spots on the steel due to a combination of both forces. In order to solve this process design issue, the impact of each force in the process will be evaluated. Evaluation is carried out using experimental data and numerical simulation.

1. Introduction

The passenger car represents the biggest stand-alone consumer of energy as well as CO2 emitter in the European Union. The EU-directive 333/2014, based on 443/2009 has set a goal of CO2 emission reduction of new passenger cars to 95 g CO2/km till 2020 [1–3]. In addition the demand for lightweight construction is strongly increasing due to range extension in electro-mobility [4].

The body in white - with 40 % of total car weight - represents a crucial adjusting screw in order to obtain these goals on time. The manufacturing method of hot forming enables the production of high-strength and complex shaped parts for automotive BIW-construction and therefore reveals a huge potential in lightweight design with over 600 million hot formed parts per year and an increasing usage in future (Fig. 1).

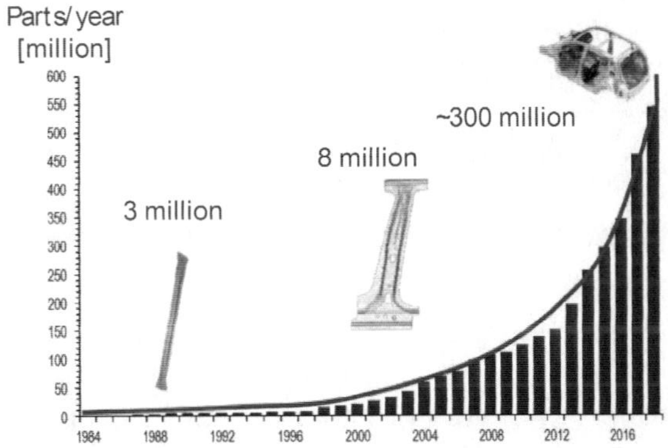

Figure 1. Development and trend of hot forming [5].

Content from this work may be used under the terms of the Creative Commons Attribution 3.0 licence. Any further distribution of this work must maintain attribution to the author(s) and the title of the work, journal citation and DOI.
Published under licence by IOP Publishing Ltd

During hot forming a boron-manganese steel alloy exhibiting a high hardenability is heated and austenitized at temperatures above 900°C to yield a fully austenitic microstructure. After a dwell time of 4-6 minutes in a gas heated roller hearth furnace (30-40 m) the blank is transferred into a press, where it is quenched and formed simultaneously.
Coating of the blanks with an Al-Si-layer is necessary in order to reduce the wear of hot forming tools due to scale formation in the furnace. Scale in hot forming tools is responsible for up to 40 % of tool wear [6]. In addition the Al-Fe-Si serves as an anti-corrosion barrier in the finished part without downstreamed shot blasting procedure.
The long roller hearth furnaces, long dwell times and high process temperatures lead to a low efficiency of the hot forming process that could be counteracted by the application of e.g. fast heating methods like induction. Using a fast inductive pre-heating to Curie-temperature of the substrate material, the necessary heating time to dwell temperature could be reduced by up to 50 % [7]. The application of fast heating in hot forming is restrained by AlSi-coating melting and dislocation due to large thermic and magnetic forces accompanied by temperature and field density gradients [8].
It remains unclear whether thermal gradients or variations in magnetic pressure are responsible for AlSi-coating dislocation during rapid heating. Therefore the thermal as well as magnetic contributions are simulated, discussed and validated with real experiments to characterize the main driving force.

2. Experiments

Experimental observations of longitudinal inductive rapid heating show a significant wave-shaped coating flow along the melting-zone of AlSi-coated hot forming steel. The temperature distribution inside of the inductor show an outwards orientated decrease in temperature with a maximum in the middle that goes hand in hand with a gradient in the magnetic field density inside of the inductor. Therefore the melting first occurs in the center-region and the flow is directed to out regions exhibiting lower temperatures and lower magnetic pressure. The experiments are conducted with an inductive heating unit shown in Fig. 2, using a capacity of 4 x 330 nF, a current of 68 A and a AC-frequency of 250 kHz. The inductor exhibits six water cooled copper coils with an internal spacing of 20 mm.

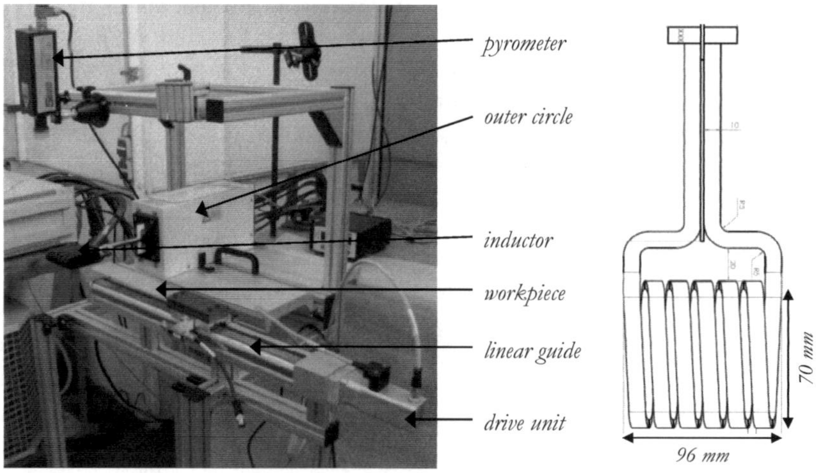

Figure 2. Experimental Setup for inductive heating experiments in dynamic and static mode.

Coated samples with geometry of 250 x 50 x 1.8 mm are inductively heated to temperatures of 700, 800 and 900°C within eight seconds. The workpiece is mounted statically inside the inductor that exhibits coil dimensions of 96 x 70 x 20 mm. The temperature is measured optically by pyrometer after conducting a calibration curve with thermocouples welded onto the material (Fig. 3, left). The highest temperature is obtained directly in the center of the inductor (~ point 2) and the temperature decreases by approximately 150 K in a 20 mm distance (~ point 4) for the 700°C measurement. Coating melting is initiated in the center after locally overshooting the melting temperature of ~580°C. Afterwards the liquid phase flows to the outer sides that subsequently undergo a phase transition. The coating thickness is measured by metallographic cross section polishes because nondestructive measurements are strongly afflicted by errors due to different intermetallic phases. Hereby the coating thickness is measured from the substrate to the coating surface and includes the intermetallic phase (with higher melting point). The thickness measurement demonstrated in Fig. 3 on the right shows the formation of a minimum/depression in the region of the intermetallic phase thickness between 10 and 14 µm and the formation of a maximum in the outer inductor regions. The coating piles up to a thickness of over 300 µm at the resolidfied wave front.

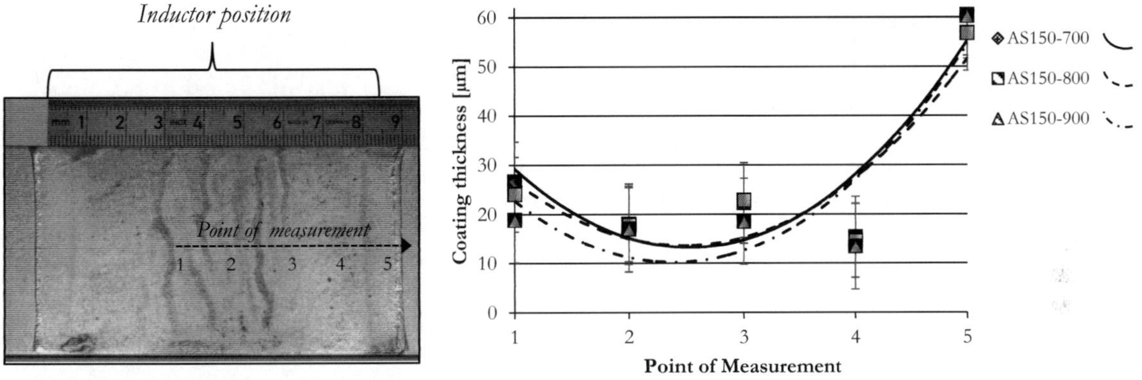

Figure 3. Positions of metallographic thickness measurements (left) and resulting coating thickness after inductive heating of AS150-coated blanks to temp. between 700 and 900°C (right).

The coating development and dislocation is equal for both sides of the blanks, therefore the gravitational effect does not play a crucial role in the phenomenon.

3. Physical model and numerical simulation

To further observe the phenomenon a CFD-Simulation using the VOF-model is built in ANSYS Fluent that only contributes to surface tension related Marangoni as well as gravitational forces. The steel substrate is treated as a non-slip boundary condition with an applied temperature field. Initial phase fraction distribution of AlSi, air and the transition zone of both phases, which results from the VoF-model, is shown in Fig. 4. All material parameters are chosen temperature depended and triangular temperature profile has been used that exhibits the highest temperature in the center of the blank and declines with 7.5 K/mm to the outer side of the blank. To reduce the necessary time for calculation a symmetry condition is used in the middle and only one half of the induction field is observed.

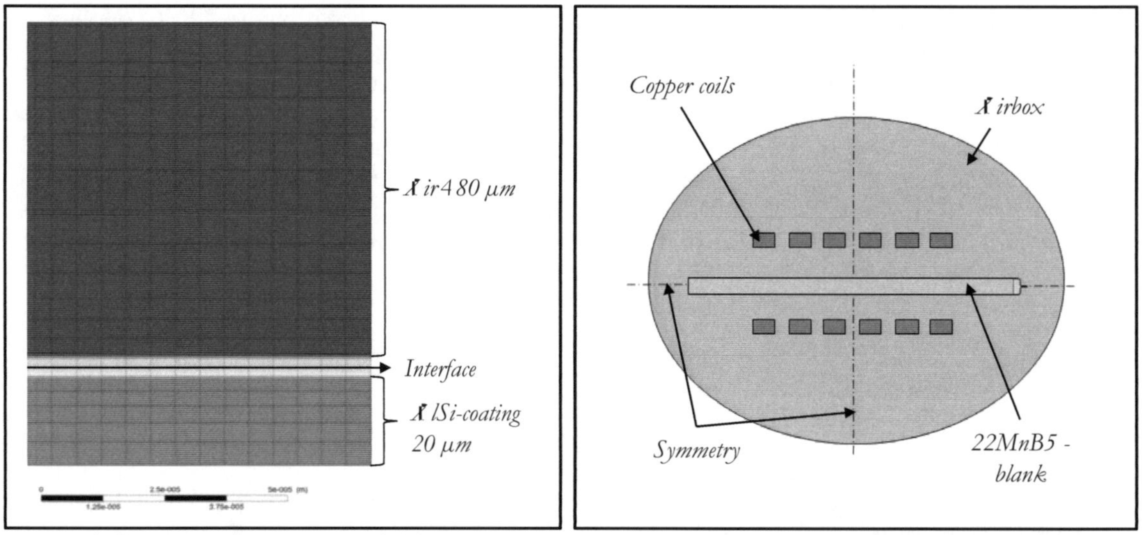

Figure 4. (left) 2D-Replacement model for VOF-Simulation containing mesh information and position of VOF 0.5-interface; (right) sketch of 2D Simulation model for coupled electromagnetic and thermal simulation.

A variable time step is used to account for a correct tracing of the interface by the PLIC-scheme. The melting is simulated with the “ANSYS melting and solidification model”. The triangular temperature profile leads to the formation of a dryspot in the center at the point of the highest temperature and a wave formation being upstream of the film delamination. If only the Marangoni force in terms of temperature gradient over the surface is taken into account – a comparable effect to the experiment can be observed in the simulation (Fig. 5).

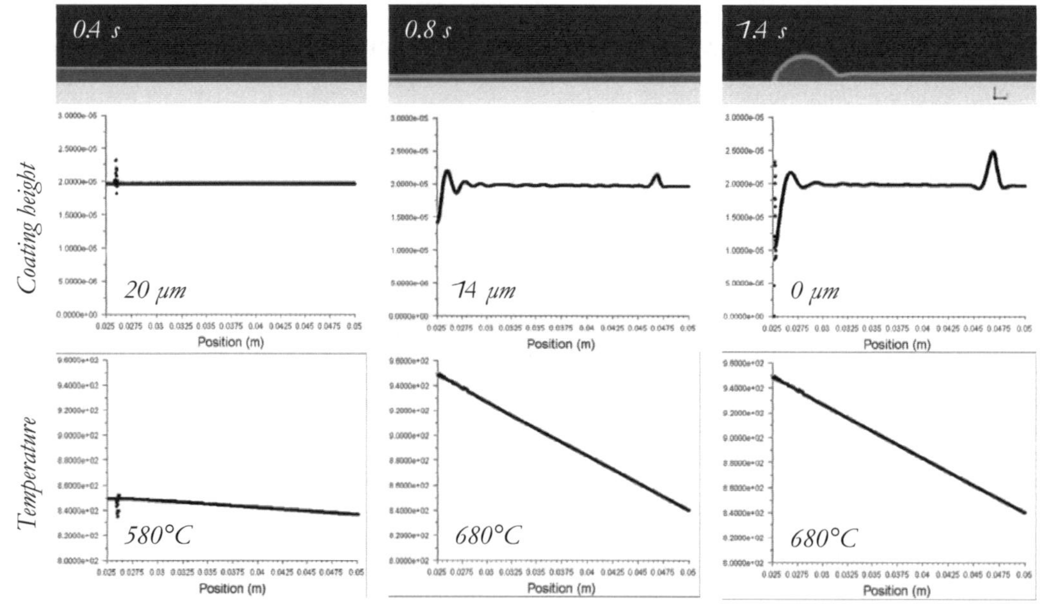

Figure 5. 2D-VOF-simulation sequence from coating melting (t=0.4s) over dryspot formation (t=0.8) till wave propagation (t=1.4s) based on a triangle temperature profile and resulting Marangoni forces by applying a temperature gradient of 70 °C over a measuring length of 50 mm.

The simulations with and without gravitational body forces show a nearly identical result. Therefore gravitation can be neglected.
Lorentzian force on the coating in x-direction due to gradients in magnetic field can be written as follows:

$$\frac{\partial P_{el,m}}{\partial x} = \frac{1}{2}\mu_0(\delta_C + \frac{\delta}{2}\left[1 - e^{\left(-2\frac{\delta_{coating}}{\delta}\right)}\right])\frac{\partial H_{em}^2}{\partial x} \quad (1)$$

where δ represents the pentration depth, μ_0 magnetic field constant and $\frac{\partial H_{em}^2}{\partial x}$ the square of the magnetic field intensity. The numerical expression is yielded based on Lupi et al. [9].
In order to evaluate the impact of Lorentzian forces, a coupled electromagnetic and thermal numerical model is built and simulation is performed by using ANSYS Multiphysics software package. The numerical model is verified by comparing the results with experimental data. The geometry is a planar two-dimensional geometry, which represents the six inductor windings, as well as the manganese steel and the AlSi coating surrounded by an air box. During the experiment electrical parameters such as voltage, current and frequency have been recorded, wherein current and frequency are serving as input parameters for the simulation. Current and frequency has been measured using a Rogowski-belt, which is placed around the inductor and outputs a current proportional voltage with the identical frequency of primary current. The values of current are oscillating around 65 – 70 A RMS. Frequency changes due to temperature dependent material properties from 252 – 217 kHz. Inductor voltage is directly measured at the connections of the inductor by removing electrical insulation coating. The voltage resulting from simulation is used, as well as the temperature, as verification for the validity of the model. The results of thermal calculation are shown in Fig. 6. The non-linear permeability of manganese steel in the simulation is the most crucial material parameter. Dependency of both, temperature and magnetic field intensity has to be taken into account. The difference of the temperature between simulation and experiment is in the range of 6 % or 10 K. The difference of electric voltage in simulation and experiment is around 16 %. Due to the simplification of the model to a 2D geometry and the neglection of the inductor connection results in a lower simulated inductivity and thus to a lower calculated voltage of the inductor. All in all, experimental and numerical results are in a good agreement and magnetic field results will be used for estimation of Lorentzian force.

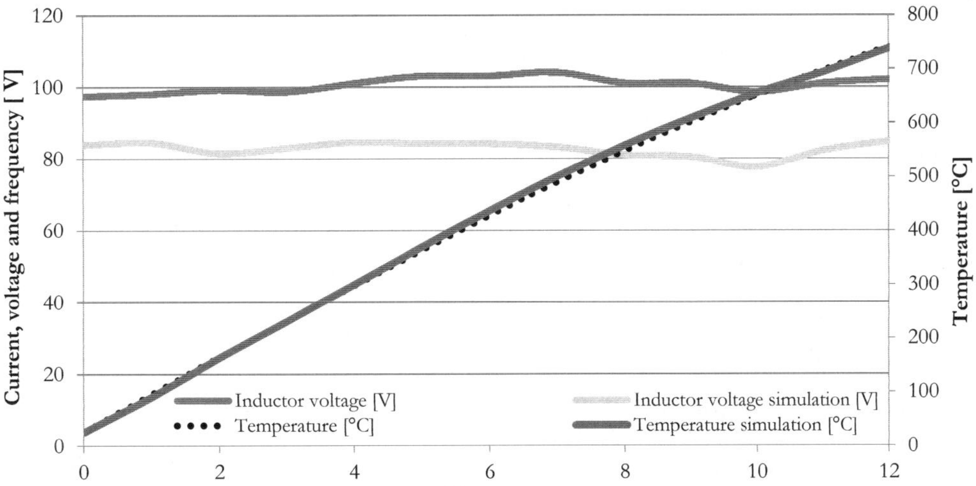

Figure 6. Measured and simulated data validation for electromagnetic force evaluation.

4. Results/Discussion

The schematic drawing in Fig. 7 gives a summary of the investigated phaenomenon. A current carrying inductor winding is located above a AlSi-coated steel substrate, that is inductively heated up due to induced eddy currents. During the heating process the AlSi coating is melted and a motion of the resulting liquid phase from the middle under the inductor to the sides can be observed. Especially in broader regions of the liquid coating, a convective flow with rolling characteristics is expected.

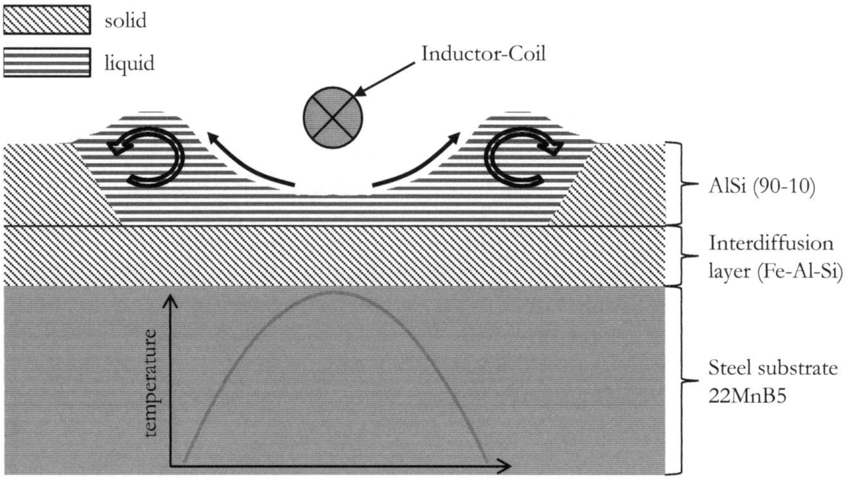

Figure 7. Schematic drawing of AlSi-coating behavior during fast and local heating above melting Temp.

The inductor creates a strong magnetic field intensity, which induces a high energy density inside of the workpiece and results in a temperature distribution with high temperature gradients. Due to these temperature gradients melting occurrence, the Marangoni force acts in the direction from high temperature to low temperature on the liquid phase. A drop in temperature is experimentally measured to approximately 150 – 200 K. Beyond the outer windings of the inductor temperature drops almost to room temperature. Especially from the middle to the outer part of the inductor high temperature gradients are created. This distribution is shown schematically in Fig. 8. According to equation (2) the temperature gradient is the driving force of the Marangoni force and is directed from high temperature to low temperature. As indicated, the equation has the dimension of a pressure. In order to obtain the driving volume force on the melt, the derivation of pressure in x-direction is calculated. σT represents the gradient of the surface tension with the temperature and is a monotonous decreasing function exhibiting a value of -3·10-4 N/mK. The data for the surface tension of AlSi(90-10) are taken out of simulations with the commercial software package JMatPro, using the TTAL8-thermotech database.

$$\frac{\partial F_\sigma}{\partial x} = P_{Marangoni} = -\sigma_T \frac{\partial T}{\partial x} \tag{2}$$

$$\frac{\partial P_{Marangoni}}{\partial x} = -\sigma_T \frac{\partial^2 T}{\partial x^2} \tag{3}$$

According to equation (1) the gradient of the squared magnetic field below the inductor and inside of the AlSi region is calculated and evaluated regarding the resulting force. In figure 8 the gradient of the squared magnetic field intensity is shown in relation to the inductor position. The highest gradient appears at the edges of the outer inductor windings, where the highest changing of potential lines of magnetic field takes place. In this case the maximum gradient is appr. $2*10^9$ [A^2/m^3], whereas the maximum magnetic field intensity appears in the middle of the inductor. As shown in Fig. 8 the maximum field intensity is appr. 6000 [A/m]. A comparison of the Lorentzian and Marangoni leads to:

$$\frac{\partial F_{el,m}}{\partial x} \Big/ \frac{\partial F_{\sigma}}{\partial x} = \frac{\mu_0 \mu_r}{2\sigma_T}\left(\delta_C + \frac{\delta}{2}\left[1 - e^{-2\left(\frac{\delta_C}{\delta}\right)}\right]\right) \frac{\partial H^2}{\partial x} \Big/ \frac{\partial T}{\partial x} \qquad (4)$$

including the following material parameters for the liquid phase: σT=3·10-4 [N/mK], penetration depth of the electromagnetic field δ=0.207 [mm], thickness of the AlSi coating δcoating=30 [µm] and relative permeability µr=1. The first part on the left side depends on coating properties only, while the second term depends on the process parameters magnetic field and temperature, which are linked directly to electric current. The first term can be calculated for given material properties and general geometry and results to $1.64*10^{-7}$. Due to small coating surface and the coating dependent properties the Marangoni forces are in this case under the inductor in a range of 20 mm around 130 times higher. Situation is of course changing with temperature equalization and location. Especially at the edges of the inductor, the Lorentzian force becomes higher compared to Marangoni forces.
Both, Marangoni forces and Lorentzian forces point in the direction of less magnetic pressure and lower temperature, which is in this situation the same direction and lead to self-strengthening effect on the coating surface.

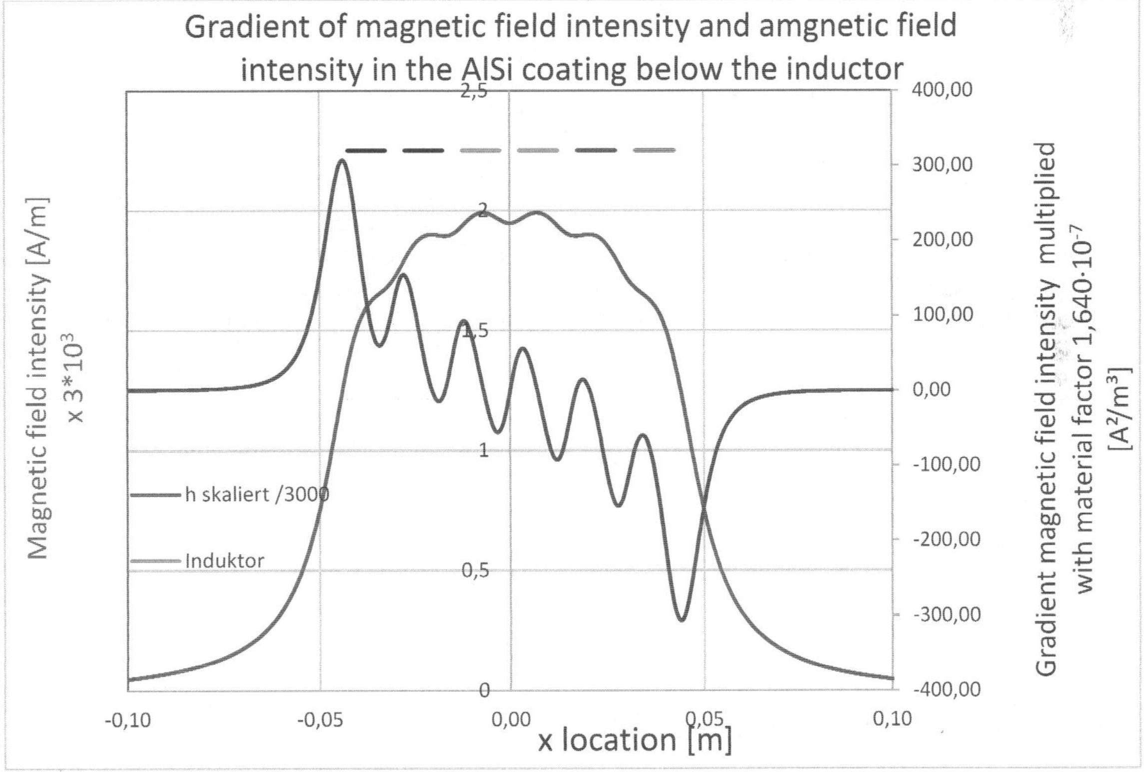

Figure 8. Resulting electromagnetic field intensity and behavior of the spatial derivative representing the electromagnetic force contribution.

5. Conclusion

The contributions of magnetic Lorentzian body forces, temperature and surface tension dependent Marangoni forces as well as gravitational forces are numerically analyzed and weighted for the present case of static inductive heating for hot forming at temperatures above 580 °C. The outcomes can be summarized as follows:

- Lorentzian forces are mainly depending on the spatial derivative of the squared magnetic field strength and pointing as a body force onto the liquid phase in direction from high magnetic to low magnetic pressures. The highest force is obtained at the outer edges of the inductor and decays in the inner areas.
- The Marangoni force is a function of the temperature gradient, acting at the rim of the melting zone. The vector of this surface tension force is congruent to the magnetic force, exhibiting a decreasing contribution in the inner inductor regions due to temperature gradient reduction.
- In summary it can be stated that the Marangoni as well as Lorentzian force contribution are adding up to the net driving force on the liquid phase with an expected higher contribution of the Marangoni forces in the present case, mainly affected by small coating thicknesses of 20-30 µm in hot forming.

Acknowledgement
The authors gratefully acknowledge the informational support of the university department "convective transport processes" of RWTH Aachen.

References

[1] Haq G, Weiss M, 2016 CO2 labelling of passenger cars in Europe: Status, challenges, and future prospects. *Energy Policy* 95, 324–335
[2] Europäisches Parlament, Amt für Veröffentlichungen, Verordnung (EU) Nr. 333/2014 des Europäischen Parlaments und des Rates vom 11. März 2014 zur Änderung der Verordnung (EG) Nr. 443/2009 hinsichtlich der Festlegung der Modalitäten für das Erreichen des Ziels für 2020 zur Verringerung der CO2-Emissionen neuer Personenkraftwagen, 2014
[3] Europäisches Parlament, Amt für Veröffentlichungen, Verordnung (EG) Nr. 443/2009 des Europäischen Parlaments und des Rates vom 23. April 2009 zur Festsetzung von Emissionsnormen für neue Personenkraftwagen im Rahmen des Gesamtkonzepts der Gemeinschaft zur Verringerung der CO2-Emissionen von Personenkraftwagen und leichten Nutzfahrzeugen 2009
[4] Hirsch J, 2016 May 11Approach for an all-aluminium lightweight car body for electric vehicles. *Materials in Car Body Engineering*, Bad Nauheim
[5] Aspacher J, Higher efficiency, output and part quality by newest developments of press technology. Strategien des Karosseriebaus, Bad Nauheim 2016 Mar 15
[6] Sartor M, Khalil T, Schoppe J, September 2012 Verringerung der Zunderbildung durch Beschichtungen. *Schmiedejournal*, 50–52.
[7] Vibrans T, 2016 Entwicklung einer Anlage zur induktiven Erwärmung von Formplatinen. elektrowärme international 2016, 39–44
[8] Opitz T, Induktive Schnellerwärmung von Formplatinen für die Warmumformung im Karosseriebau. HTM J. *Heat Treatent Mat.* 70 2015, 201–206
[9] Lupi S, Forzan M et al, 2015 Induction and direct resistance heating: Theory and numerical modeling, Springer, Cham

IOP Conf. Series: Materials Science and Engineering **355** (2018) 012014 doi:10.1088/1757-899X/355/1/012014

Numerical investigation of effects on blanks for press hardening process during longitudinal flux heating

André Dietrich and Bernard Nacke

Institute of Electrotechnology, Leibniz Universität Hannover, Wilhelm-Busch-Str. 4, 30167 Hannover, Germany

E-mail: dietrich@etp.uni-hannover.de, nacke@etp.uni-hannover.de

Abstract. With the induction heating technology, it is possible to heat up blanks for the press hardening process in 20 s or less. Furthermore, the dimension of an induction system is small and easy to control in comparison to conventional heating systems. To bring the induction heating technology to warm forming industry it is necessary to analyze the process under the view of induction. This paper investigates the edge- and end-effects of a batch heated blank. The results facilitate the later design of induction heating systems for the batch process.

1. Introduction

In the last years different induction heating concepts for blanks were tested. The article [1] analyzes the potential of continuous heating with longitudinal and transversal flux concepts. With those heating concepts it is possible to heat blanks up to the Curie temperature. The potential to heat blanks above the Curie temperature was described in [2] using a single side induction coil and in [3] using the concept of a single side induction heater too and examine the influence of a longitudinal flux heating coil on the temperature distribution. The influence of longitudinal flux heating on holes and block-outs was investigated in [4] and [5]. This paper analyzes an induction heating system which allows temperatures up to 950 °C. The advantage of single stage induction heating is that conventional heating systems like a gas furnace aren't necessary if a homogeneous final temperature distribution of the blank can be reached only by induction heating. In the hot sheet metal forming process blanks are heated up to a temperature of 930 °C to 950 °C for uncoated material and AlSi coated resp. 890 °C to 920 °C for Zn coated material in order to reach the austenitisation temperature and compensate the thermal losses, which occur during the transport from the heating system to the press.

2. Numerical model

For the numerical modelling and calculation, the commercial software package ANSYS 16.2 was used, which bases on the finite-element-method (FEM). The calculation algorithm which is used in the study is divided in two sub-steps. In each iteration step a harmonic electromagnetic and a transient thermal calculation are performed. To solve the Maxwell Equation's for the harmonic electromagnetic field the magnetic vector potential is used. The result of this solution is the heat source distribution in the work piece. In the second sub-step the heat transfer equation is solved for a time interval Δt taking into account thermal losses due to radiation and convection. The solution of the electromagnetic calculation is the input value for the heat transfer equation.

Content from this work may be used under the terms of the Creative Commons Attribution 3.0 licence. Any further distribution of this work must maintain attribution to the author(s) and the title of the work, journal citation and DOI.
Published under licence by IOP Publishing Ltd

The output value is the temperature distribution. In the next iteration step the temperature distribution is the input value for a new harmonic electromagnetic calculation with adjusted material properties like specific electrical conductivity, relative permeability, specific thermal conductivity and the specific heat capacity. For the calculation of the following results a verified 3-dimensional model is used [6]. Due to the symmetry only the half of the system is modelled in order to reduce the calculation time. The induction coil has a rectangular cross section (Figure 1). For an exact calculation it is important to mesh the edges of the blank very fine. The calculations are based on the material properties of 22MnB5 steel [7]. The reference thickness of the blank is 1.5 mm, because in many industrial application blanks with a thickness in this range are used. In most cases the width and the length of the blank is 500 mm. The coil has a number of 8 windings. The width of one winding is 60 mm and the distance between the windings is 5 mm. The frequency of the inductor current is 400 kHz. The air-gap in vertical direction is 40 mm and in horizontal direction 50 mm on each side.

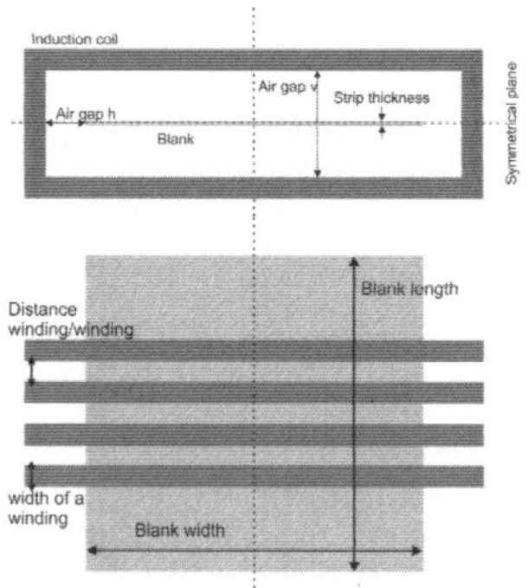

Figure 1. Sketch of the numerical model

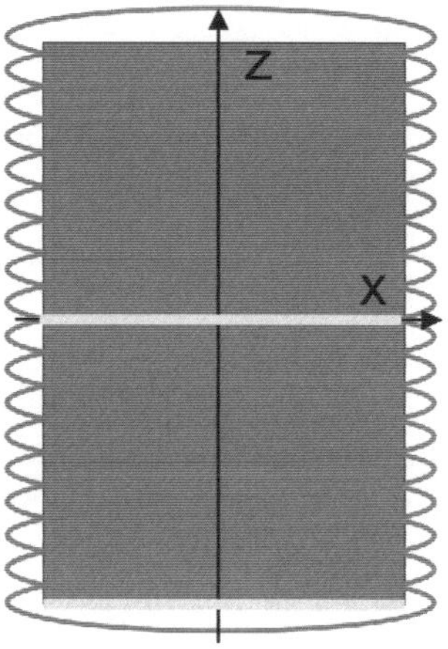

Figure 2. Evaluation path in x-direction

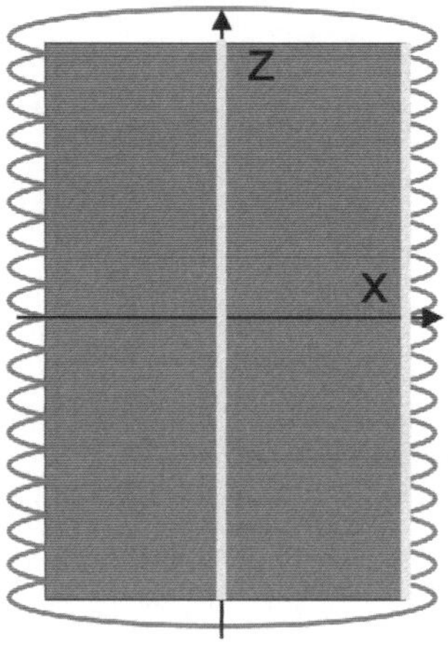

Figure 3. Evaluation path in z-direction

3. Numerical investigation

3.1. Influence of the horizontal air gap on the temperature distribution

The air gap is an important geometrical value in the induction heating process. It acts on the temperature distribution and the electrical efficiency. First the modification of the air gap in horizontal direction is analyzed. This is the distance between the edge of the blank and the normal part of the induction coil. The temperature is evaluated in two directions and each on two paths (Figure 2 and 3). In figure 4 and figure 5 the temperature profiles parallel to the windings are shown. Between 0.05 m and 0.45 m the temperature is nearly homogeneous.

Close to the edge the blank is overheated. The margin of the overheating depends of the size of the horizontal air gap. The overheating at the edge falls with increasing the horizontal air gap. At the end of the blank (Figure 5) the temperature profile is qualitatively the same but around 2 percent lower because the magnetic field spreads at the end of the induction coil.

In figure 6 and figure 7 the temperature profile normal to the windings (Figure 3) is illustrated. The temperature profile in figure 6 is evaluated at the position $x = 0$ m. For the four air gaps the temperature has qualitatively the same trend. But in figure 7 the temperature along the edge is different for the air gaps. If the horizontal air gap is small the magnetic field which is generated from the normal part of the induction coil has more influence on the

IOP Conf. Series: Materials Science and Engineering **355** (2018) 012014 doi:10.1088/1757-899X/355/1/012014

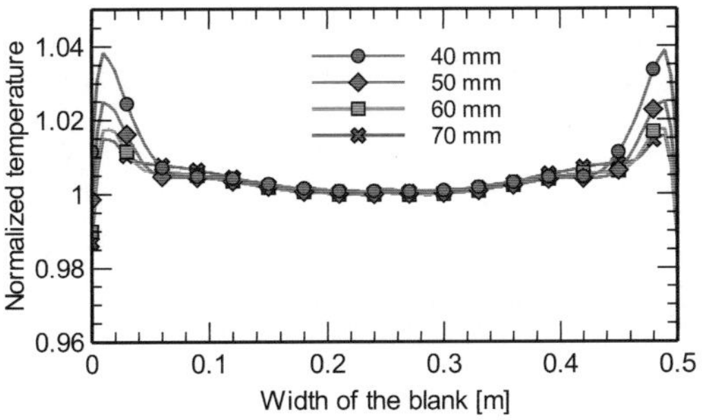

Figure 4. Temperature in x-direction in the centre

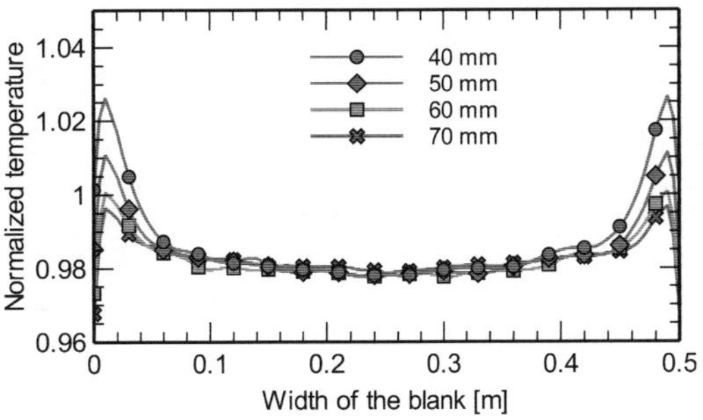

Figure 5. Temperature in x-direction at the end

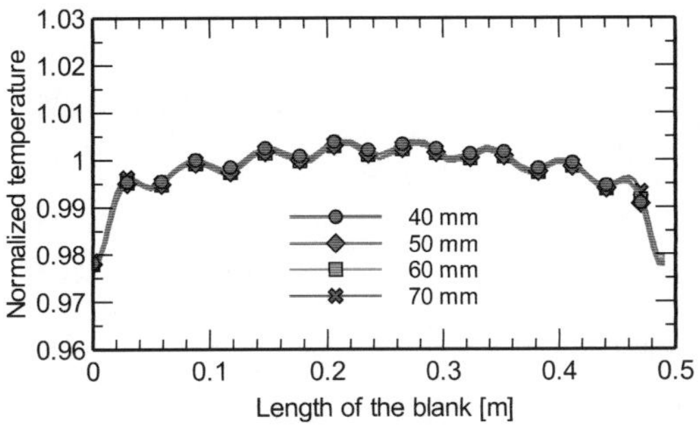

Figure 6. Temperature in z-direction in the centre

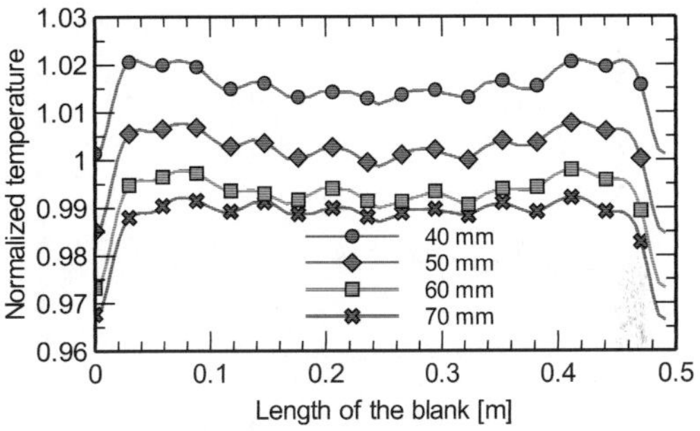

Figure 7. Temperature in z-direction at the edge

temperature of the edge and causes a higher value (Figure 7). The ripples in figure 6 and figure 7 are produced by the distance between the windings. A part of the magnetic field doesn't close over the whole coil rather over a single winding. That means the magnetic field between the windings is lower than under a winding and therefore the temperature too.

3.2. Influence of the vertical air gap on the temperature distribution

The next analyzed parameter is the vertical air gap. The temperature profile at position $z = 0$ mm is influenced by the vertical air gap (Figure 8). With increasing the air gap the homogeneous zone gets smaller and the shape of the curve is smoother. Due to the increased distance between the blank and the horizontal part of the induction coil the coupling becomes weaker. The influence of the normal part of the induction coil increases and, therefore, the magnetic field is greater in the edge area. The temperature at the end of the blank is shown in figure 9. The increasing of the vertical air gap results in a lower temperature at the end of the blank because of a weaker magnetic field in this area but the difference from minimum to maximum is the same for each curve.

In figure 10 and figure 11 the temperature along the z-axis is illustrated. With a greater air gap the temperature at the end of the blank decreases. Furthermore, the ripples of the graphs are reduced because the influence of the magnetic field around a single winding decreases with a greater air gap and therefore, the magnetic field is more homogeneous.

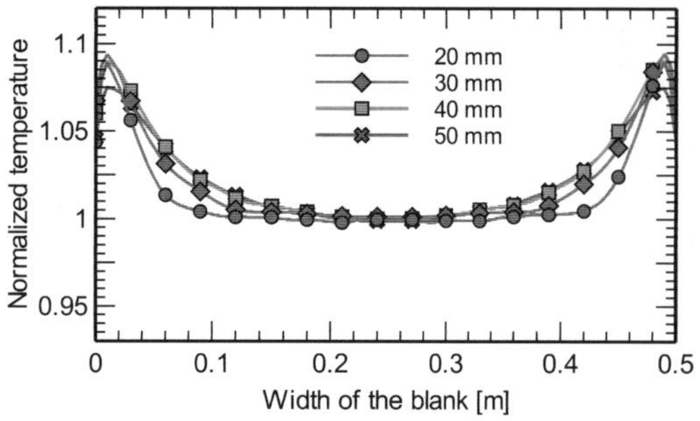

Figure 8. Temperature in x-direction in the centre

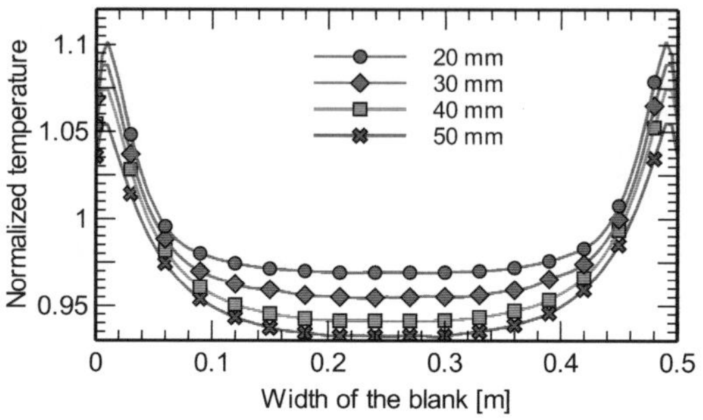

Figure 9. Temperature in x-direction at the end

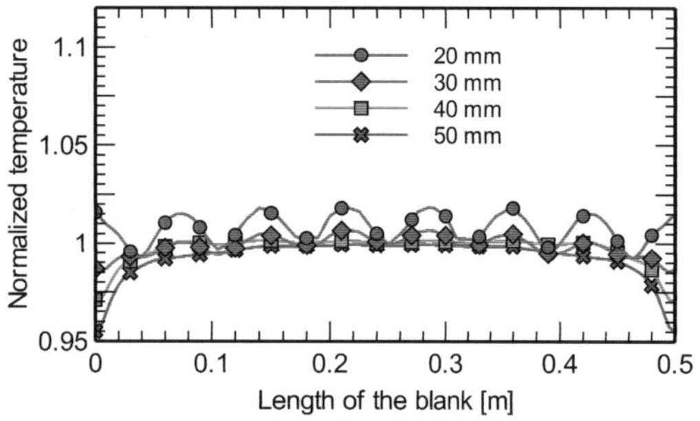

Figure 10. Temperature in z-direction in the centre

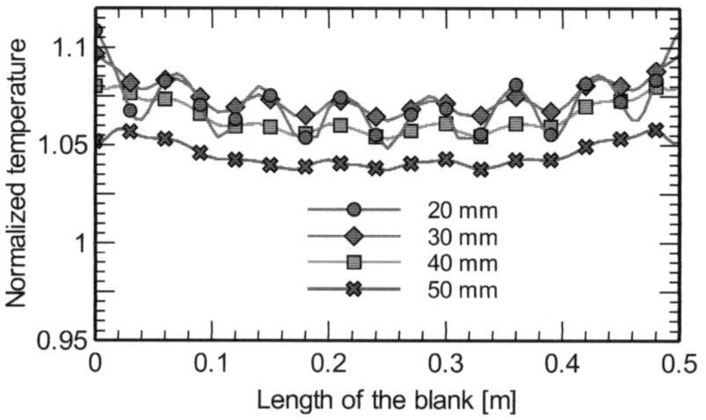

Figure 11. Temperature in z-direction at the edge

3.3. Influence of the length of the blank on the temperature distribution

The next parameter in this investigation is the length of the blank. Blanks in the press hardening process are different in length. So it is important to know in with range the length of a blank can be varying for a nearly homogeneous heating. The modification of the length has no influence on the temperature in x-direction at position $z = 0$ m (Figure 12). But at the end of the blank there is a dependence of the length on the temperature profile (Figure 13). If the blank is longer than 480 mm the temperature difference between the position $z = 0$ m and the end of the blank is more than 1 percent. If the blank is shorter than 480 mm the temperature profile is no longer influenced that means the temperature is nearly the same as at position $z = 0$ m.

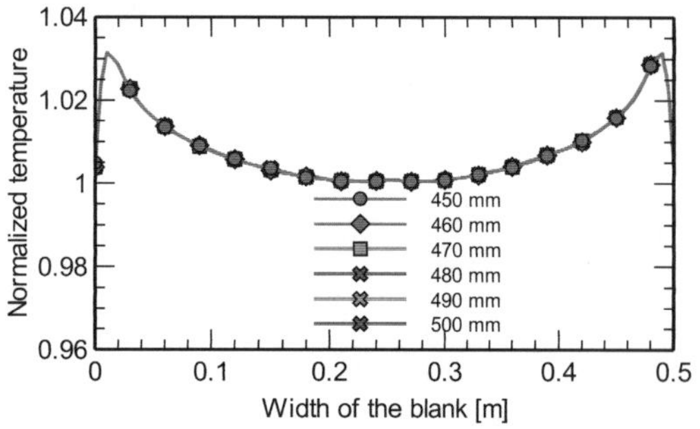

Figure 12. Temperature in x-direction in the centre

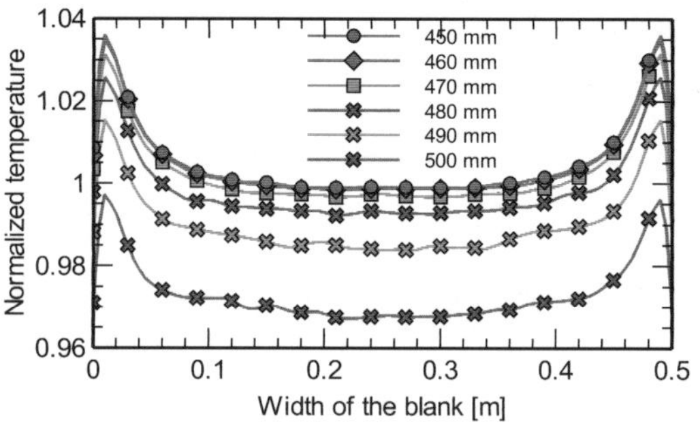

Figure 13. Temperature in x-direction at the end

The temperature profile in z-direction for position $x = 0$ m and at the edge of blank is shown in figure 14 and figure 15. At position $x = 0$ m (Figure 14) the temperature is very homogeneous for a length of 480 mm or shorter. The temperature directly at the edge is around 1 percent higher than in the centre of the blank (Figure 15). For a longer blank the temperature drops because of a spread magnetic field at the end of the coil. The temperature profiles of the curves are qualitatively the same and the difference between minimum and maximum for each curve are 2.9 percent.

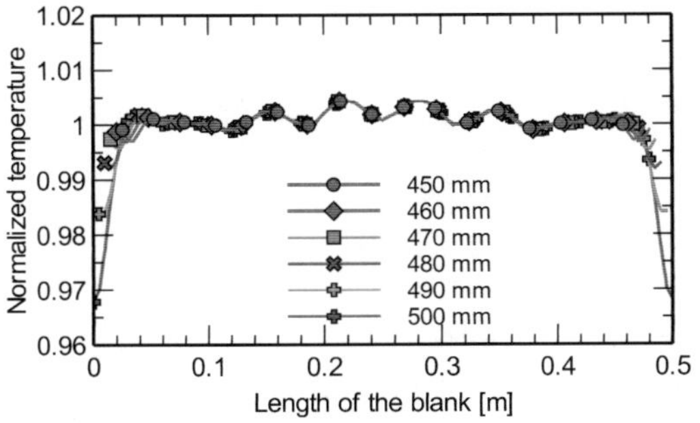

Figure 14. Temperature in z-direction in the centre

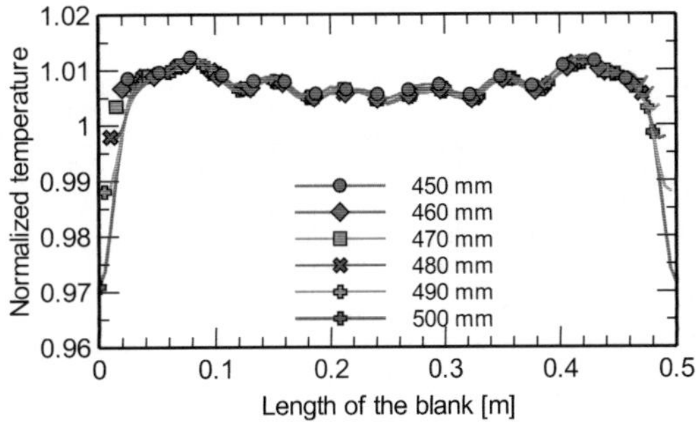

Figure 15. Temperature in z-direction at the edge

3.4. Influence of the thickness of the blank on the temperature distribution

The thickness of the blanks which are used in the press hardening process varies in a range of 1 mm to 3 mm. Between 0.06 m and 0.44 m the difference in the temperature profile (Figure 16 and Figure 17) isn't big but near the edge of the blank the temperature increases with the thickness of the blank. With a thicker blank more Joule Heat is generated in the vertical part at the edge because the way for the electrical current gets longer. This Joule Heat superposes with the Joule Heat of the surface and results in higher temperatures at the edge.

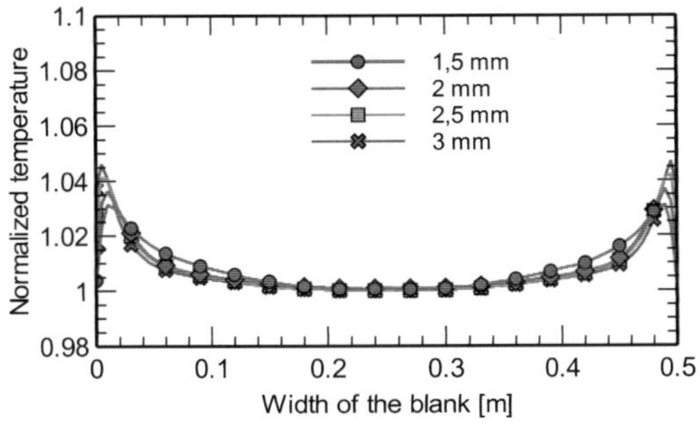

Figure 16. Temperature in x-direction in the centre

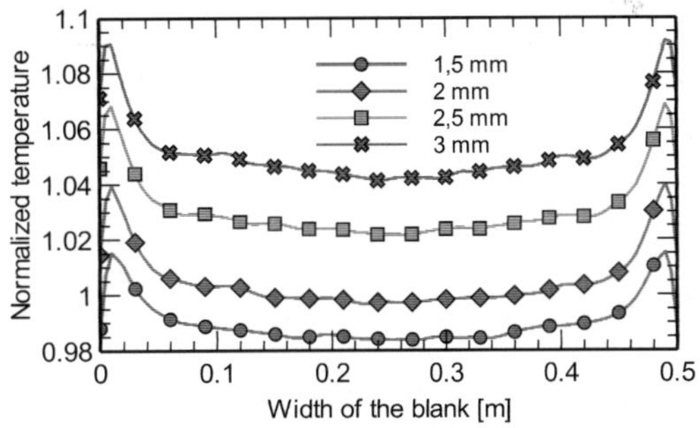

Figure 17. Temperature in x-direction at the end

Figure 18 and figure 19 show the temperature profiles in z-direction. The temperature profiles in the centre (Figure 18) have nearly the same trend for the different blank thicknesses. Only at the end of the blank the temperature rises with thicker blanks.

At the edge of the blank (Fig. 19) the temperature is higher than in the centre. Furthermore, the temperature of thicker blanks is higher than for smaller ones at the edge of the blank.

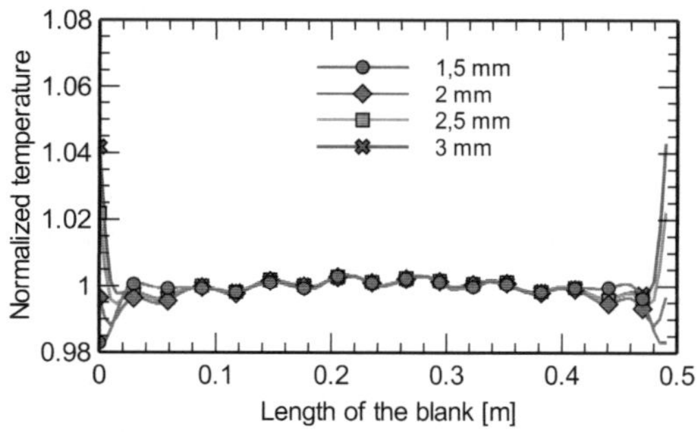

Figure 18. Temperature in z-direction in the centre

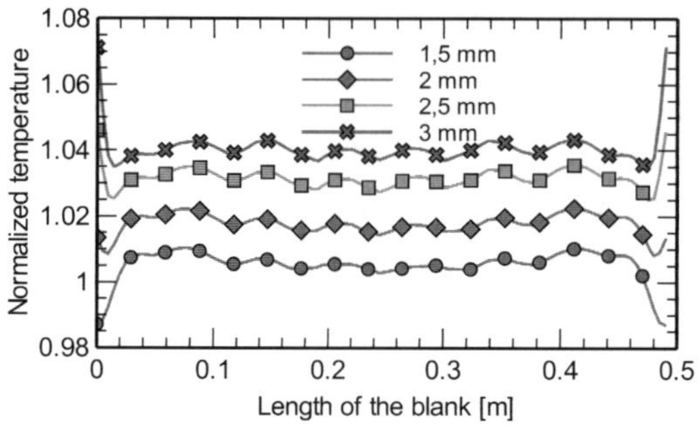

Figure 19. Temperature in z-direction at the edge

4. Conclusion

With a 3D numerical model, the batch heating process for thin blanks in longitudinal flux field was investigated. With increasing the horizontal air gap the overheating at the edge and at the end of the blank could be reduced. The vertical air gap has only a small influence on the temperature in the centre of the blank but at the end the temperature drops around 5 percent for a large air gap. The length of a blank influences only the temperature at the ends. For the investigated length the temperature drops 3 percent. The last analyzed parameter was the thickness of the blank. With increasing the thickness, the temperature at the edge and end increases. The influence of the thickness on the temperature distribution at the end of the blank has to be investigated in more detail. In the future an experimental setup is planed for heating a b-pillar.

Acknowledgments

The research project Series-production of inductively heated blanks for hot sheet metal forming (IFG-No. 18738 N) from the Research Association for steel Application (FOSTA) and the Research Association of Automotive Technology (FAT) is supported by the Federal Ministry of Economic Affairs and Energy through the German Federation of Industrial Research Associations (AiF) as part of the program for promoting industrial cooperative research (IGF) on the basis of a decision by the German Bundestag.

References

[1] Opitz T and Vibrans T 2015 Induktive Schnellerwärmung von Formplatinen für die Warmumformung im Karosseriebau *HTM J. Heat Treatm. Mat.* **70** **6** 261-6

[2] Tröster T and Niewel J 2014 *Inductive heating of blanks and determination of corresponding process windows for press hardening* (Düsseldorf) FOSTA Research Association for Steel Application vol 805

[3] Todzy T 2015 *Prozessfenster beim Presshärten bei schneller Erwärmung von Stahlplatinen mit Aluminium-Silizium-Beschichtung* (Aachen: Shaker-Verlag)

[4] Dietrich A Nacke B Pfeifer F Marten T and Tröster T 2017 Investigation of geometrical discontinuities in blanks for hot sheet metal forming process under the influence of induction heating *5th International Conference on Steels in Cars and Trucks* Noordwijkerhout/Amsterdam

[5] Dietrich A Nacke B Pfeifer F Marten T and Tröster T 2017 Effects of Holes in Blanks for Press Hardening Process due to Induction Heating *XVIII International UIE-Congress Electrotechnologies for Material Processing* Hannover pp 327-32

[6] Dietrich A Schülbe H and Nacke B 2016 Induktive Längsfelderwärmung von dünnem Metallband: numerische Berechnung und experimentelle Verifikation *Elektrowärme International* **74** 65-70

[7] Vibrans T 2016 *Induktive Erwärmung von Formplatinen für die Warmumformung* Technische Universität Chemnitz

IOP Conf. Series: Materials Science and Engineering **355** (2018) 012015 doi:10.1088/1757-899X/355/1/012015

Comparison of single and consecutive dual frequency induction surface hardening of gear wheels

J Barglik[1], K Ducki[1], D Kukla[2], J Mizera[2], G Mrówka-Nowotnik[3], J Sieniawski[3] and A Smalcerz[1]

[1]Silesian University of Technology, ul. Krasińskiego 8, 40-019 Katowice, Poland,
[2]Warsaw University of Technology, ul. Wołoska 141, 02-507 Warszawa, Poland,
[3]Rzeszów Technical University, ul. Żwirki i Wigury 4, 35-959 Rzeszów, Poland

jerzy.barglik@polsl.pl, jmizera@inmat.pw.edu.pl, jansien@prz.edu.pl

Abstract. Mathematical modelling of single and consecutive dual - frequency induction surface hardening systems are presented and compared. The both models are solved by the 3D FEM-based professional software supported by a number of own numerical procedures. The methodology is illustrated with some examples of surface induction hardening of a gear wheel made of steel 41Cr4. The computations are in a good accordance with experiments provided on the laboratory stand.

1. Introduction

Induction hardening is a kind of heat treatment where a steel body or its selected part is heated by induction to an appropriate temperature and then immediately quenched [1]. The surface induction hardening is connected with hardening of a thin surface layer only and keeping soft the internal part of the treated material. In general, the induction surface hardening consists of two consecutive stages with a short austenitisation break between them. The time evolutions of the process are presented in Fig.1 for the single frequency induction hardening (SFIH) (on left) and for the consecutive dual frequency induction hardening (CDFIH) (on right).

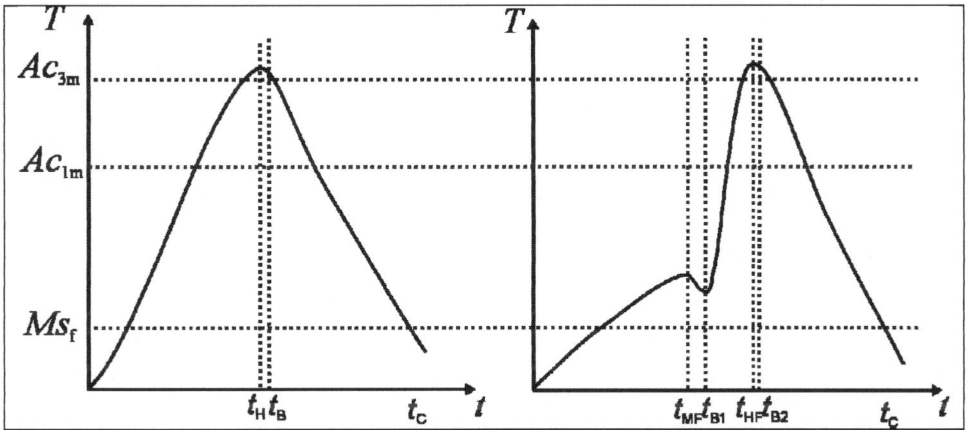

Figure 1. Time evolution of temperature during single (left) and dual frequency induction surface hardening (right).

Content from this work may be used under the terms of the Creative Commons Attribution 3.0 licence. Any further distribution of this work must maintain attribution to the author(s) and the title of the work, journal citation and DOI.
Published under licence by IOP Publishing Ltd

The first stage of the SFIH process is the fast induction heating. In a thin surface layer the treated material reaches the hardening temperature T_h being bigger than the modified upper critical temperature Ac_{3m} [2].

$$T\big|_{t=t_B} \geq Ac_{3m} \quad (1)$$

For deeper layers the hardened body reaches temperatures bigger than the modified lower critical temperature Ac_{1m}, but smaller than the modified upper critical temperature Ac_{3m} and the microstructure contains not only austenite, but also pearlite and carbides.

$$Ac_{3m} \geq T\big|_{t=t_B} \geq Ac_{1m} \quad (2)$$

Finally, for the internal part of the body the material does not contain austenite at all.

$$T\big|_{t=t_B} \leq Ac_{1m} \quad (3)$$

For the dual-frequency process, the induction heating could be realized as simultaneous or consecutive. In the latter case the heating is realized in two consecutive time steps t_{MF} and t_{HF} respectively with a short break between them. The first step is the medium frequency (MF) induction heating. Then almost immediately (short break time t_{B2}) the high frequency (HF) induction heating is realized. The heating terminates when the average temperature in the hardened zone exceeds Ac_{3m}. However it depends on velocity of heating. In order to determine it the Time-Temperature-Austenitisation (TTA) diagram for the investigated steel 41Cr4 is measured (Fig. 2).

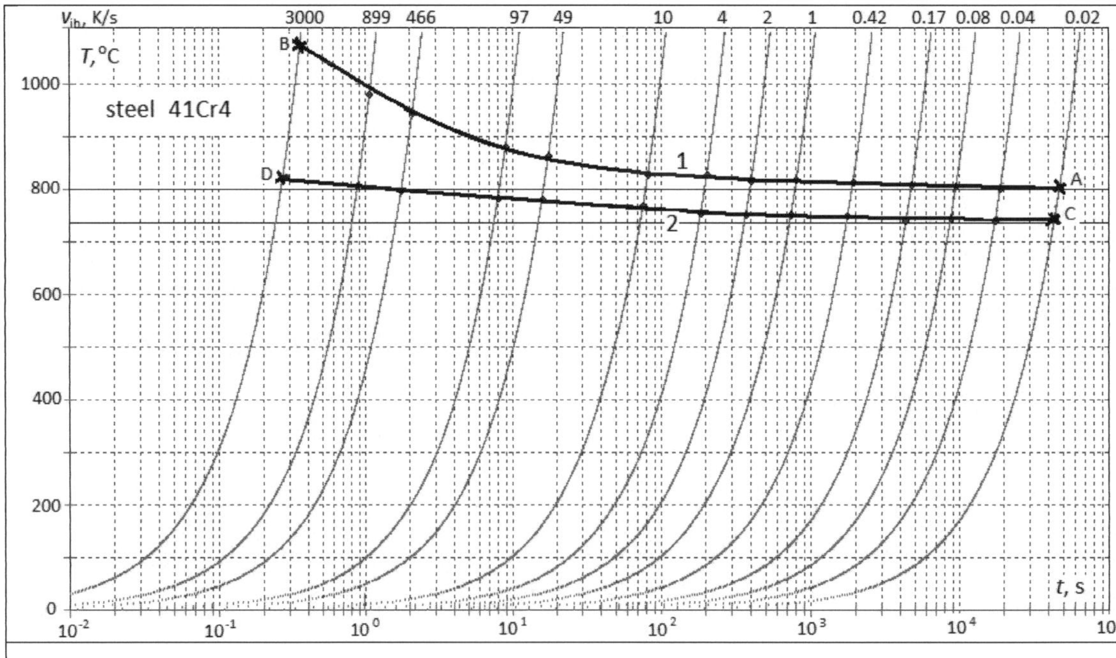

Figure 2. TTA diagram for steel 41Cr4.

Curve 1 in Fig. 2 represents dependence of Ac_{3m} on velocity of induction heating v_{ih}. For slow heating Ac_3 = 801^{0}C (point A at the curve 1). For very fast heating the modified upper criterial temperature Ac_{3m} is distinctly bigger. For velocity of heating v_{ih} = 3000 K/s Ac_{3m} = 1075^{0}C (point B at the curve 1). Curve 2 in Fig. 2 represents dependence of Ac_{1m} on velocity of induction heating v_{ih}. For slow heating Ac_1 = 736^{0}C (point C at the curve 2). For very fast heating the modified lower criterial temperature Ac_{1m} is bigger. For v_{ih} = 3000 K/s Ac_{1m} = 815^{0}C (point D at the curve 2). The second stage – quenching – makes it possible to achieve the requested martensitic microstructure [3]. The resultant hardness depends on the speed of cooling. The real character of this dependence can be determined from the Continuous-Cooling-Transformation (CCT) diagram. For steel 41Cr4, in order to obtain uniform martensitic microstructure, the velocity of cooling should be bigger than 500 K/s and the final temperature lower than the martensite finish temperature Ms_f. During a short break between both stages the temperature of the body decreases, but at the surface layer it is still higher than Ac_{3m}.

2. Mathematical modelling

In the paper we focus our attention on the spin induction hardening, where the material is heated simultaneously by a ring inductor [4]. We use single and dual-frequency hardening systems which allows obtaining the prescribed hardness profile. From the physical viewpoint, induction heating is a strongly nonlinear multi-physic process (Fig. 3). The interaction between fields are connected mostly by means of temperature dependences of material properties.

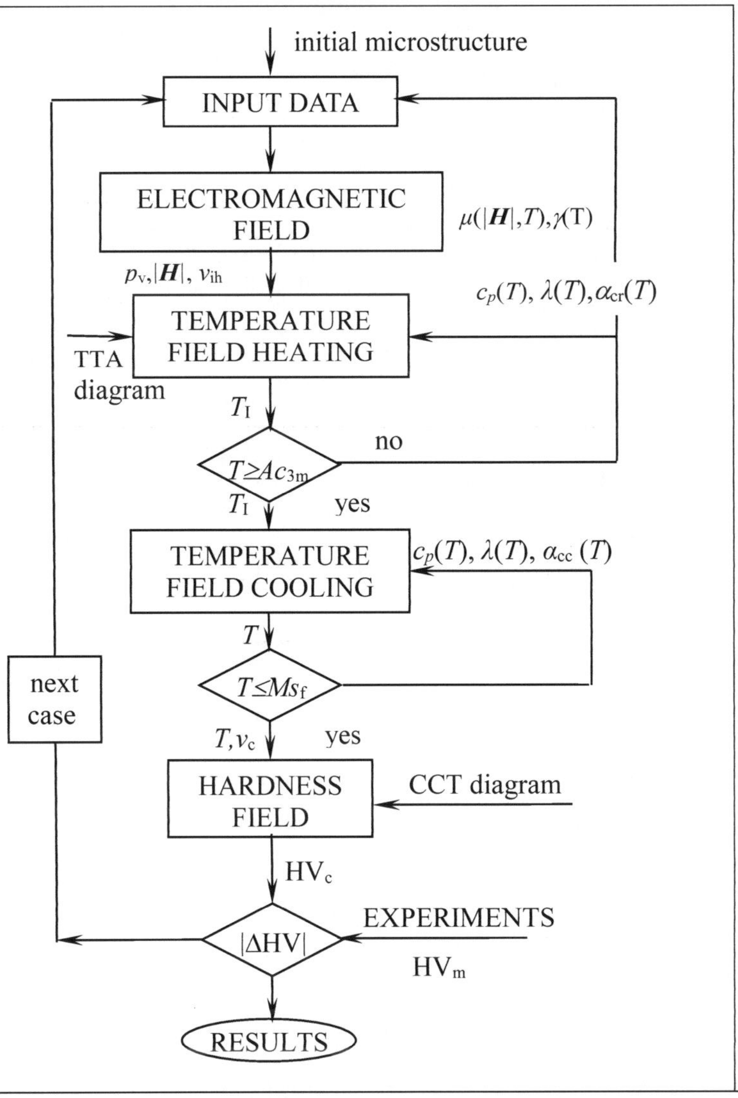

Figure 3. Algorithm of the computation.

Electromagnetic and temperature fields were analyzed by means of FEM-based software. The total volumetric power density p_v consists of two components, but the hysteresis losses are neglected:

IOP Conf. Series: Materials Science and Engineering **355** (2018) 012015 doi:10.1088/1757-899X/355/1/012015

$$p_v = p_H + p_J \tag{4}$$

$$p_H = p_H(|\boldsymbol{H}|, f) \tag{5}$$

$$p_v \approx p_J = \frac{|\boldsymbol{J}_{ind}|^2}{\gamma} \tag{6}$$

where $\boldsymbol{H}$ denotes the magnetic strength, f the frequency, J_{ind} eddy current density and γ electric conductivity.
Dependences of the specific heat c_p, the thermal conductivity λ and the electric conductivity γ on temperature are shown in Fig. 4 – 6. Dependence (7) of the relative magnetic permeability μ_r on temperature for $|\boldsymbol{H}| = \text{const}$ was shown in Fig. 7 [5].

$$\mu_r = f(|\boldsymbol{H}|, T) \tag{7}$$

The heat transfer was given by (8). Radiation was taken into account, however multiple reflection phenomena were neglected [6].

$$-\lambda \cdot \frac{\partial T}{\partial n} = \alpha_c (T - T_{ac}) + \sigma_0 \varepsilon (T^4 - T_{ar}^4) \tag{8}$$

where α_c denotes convection heat transfer coefficient, T_{ac} temperature of convection environment, σ_0 – Stefan constant, ε – emissivity, T_{ar} – temperature of radiation environment.

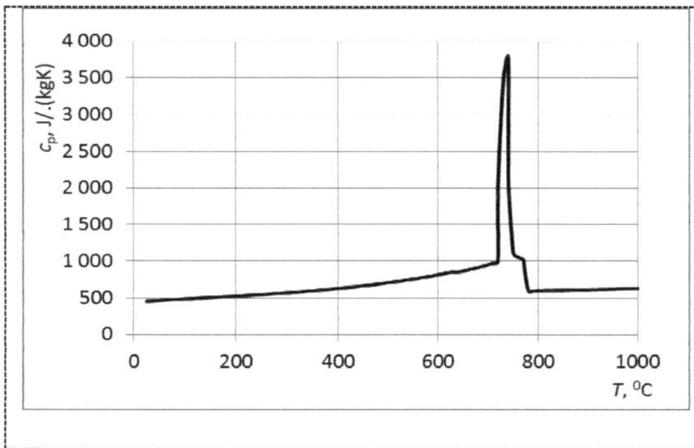

Figure 4. Dependence of the specific heat on temperature.

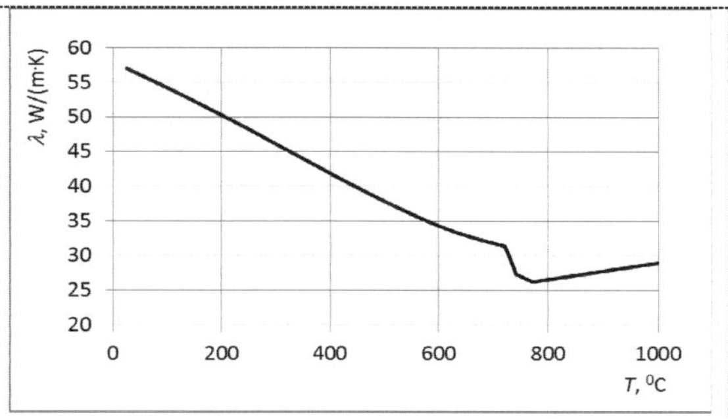

Figure 5. Dependence of the thermal conductivity of temperature.

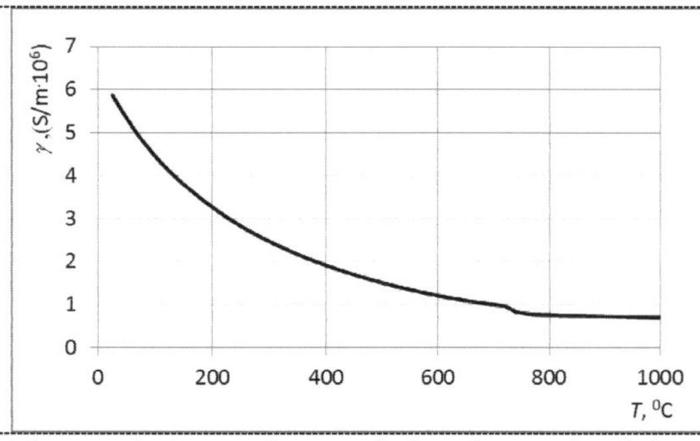

Figure 6. Dependence of electric conductivity on temperature.

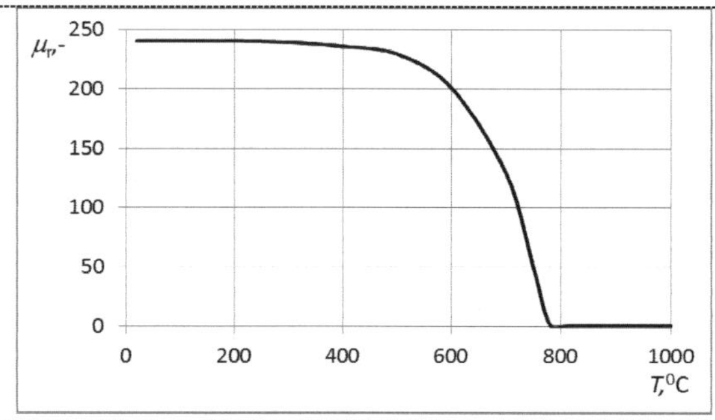

Figure 7. Dependence of the relative magnetic permeability on temperature.

IOP Conf. Series: Materials Science and Engineering **355** (2018) 012015 doi:10.1088/1757-899X/355/1/012015

The computation of temperature field for induction heating terminates when the average temperature of the thin surface zone exceeds Ac_{3m}. Then temperature field for cooling by spraying was calculated. Dependences of the specific heat c_p, the thermal conductivity λ and convection heat transfer α_{cc} on temperature were taken into account. The computation of temperature field for cooling terminates when the inequality (9) was satisfied.

$$T\big|_{t=t_C} \le Ms_f \tag{9}$$

where t_C denotes the time when cooling terminates and Ms_f means the temperature when the martensitic transformation is completed.

And finally hardness and microstructure distribution were determined and compared with measurements.

3. Illustrative Example

The numerical solution of the induction hardening of gear wheels was provided by the Flux3D software with some own procedures. An attention was paid to the convergence of results in the dependence on the density of discretization meshes for the electromagnetic and temperature fields and on the position of the external artificial boundary.

Basic parameters and dimensions of the system are as follows:

Gear wheel: teeth number n = 16, width of the tooth ring b = 0.006 m, external diameter d_e = 0.0356 m, internal diameter d_i = 0.0269 m, diameter of the hole d_h = 0.016 m, material: steel 41Cr4 (its chemical composition is presented in Tab. 1).

Table 1. Chemical composition of steel 41Cr4.

Element	C	Cr	Si	Mn	Ni	Cu	P
%	0.4	1.05	0.24	0.73	0.16	0.16	0.025

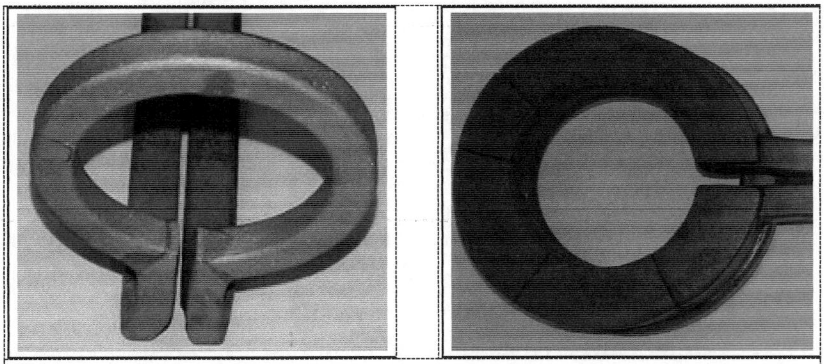

Figure 8. MF inductor (left) and HF inductor – sprayer system (right).

MF inductor (left part of Fig. 8): number of turns n = 1, external diameter d_e = 0.054 m, internal diameter d_i = 0.0395 m, height h = 0.007 m, length of bus-bars l_b = 0.363 m, distance between them Δl = 0.003 m

HF inductor (right part of Fig. 8): number of turns n = 1, external diameter d_e = 0.061 m, internal diameter d_i = 0.0395 m, height of coil h = 0.007 m total height (coil + concentrator) h_t = 0.021 m,

Flux concentrator: external diameter d_e = 0.0815 m, internal diameter d_i = 0.039 m, thickness of upper and lower layer b = 0.005 m.

Sprayer : distance between inductor and sprayer 0.02 m, external diameter d_e = 0.085 m, internal diameter d_i = 0.061 m, quenchant: 10 % of Aqua Quench 140, 90 % of water.

Heat transfer parameters: radiation for heating only, the temperature of convection and radiation environment T_{ac} = T_{ar} = 20°C, the heat transfer coefficient during heating α_{cr} = 20 W/(m^2·K),

IOP Conf. Series: Materials Science and Engineering **355** (2018) 012015 doi:10.1088/1757-899X/355/1/012015

temperature of quenchant T_q = 30^{0}C, convection heat transfer coefficient during cooling α_{cc} = 1200 W/(m^2·K),

Modified criterial temperatures for real heating conditions: heating velocity v_{ih} = 200 – 400 K/s, Ac_{3m} = 920 – 950^{0}C, Ac_{1m} = 770 –780^{0}C,

Parameters of the induction heating for various cases: medium frequency: power of the generator: P_{MF} = 60 kW, current I_{MF} = 1385 A, heating time Δt_{MF} = 4 – 7 s, frequency f = 36 kHz, high frequency: P_{HF} = 20 kW, current I_{HF} = 500 A, frequency f = 280 kHz, Δt_{HF} = 0.7 – 1,5 s, rotation velocity v_r = 2 r/s.

Below three examples of the induction hardening are presented. First the CDFIH process was analyzed for following parameters: MF heating: t_{MF} = 4 s, I_{MF} = 1385 A, f_{MF} = 36 kHz, Break: Δt_{B1} = 0.2 s, HF heating: Δt_{HF} = 0.7 s, I_{HF} = 500 A, f_{HF} = 240 kHz.

Temperature distribution within the tooth after MF heating is presented on left part of Fig. 9. For Δt_{MF} = 4 s average temperature within the tooth T_{av1} = 475^{0}C, which is sufficiently lower than Ac_{1m}. After HF heating the average temperature in the thin hardened zone along the working surface of the tooth T_{av2} = 923^{0}C.

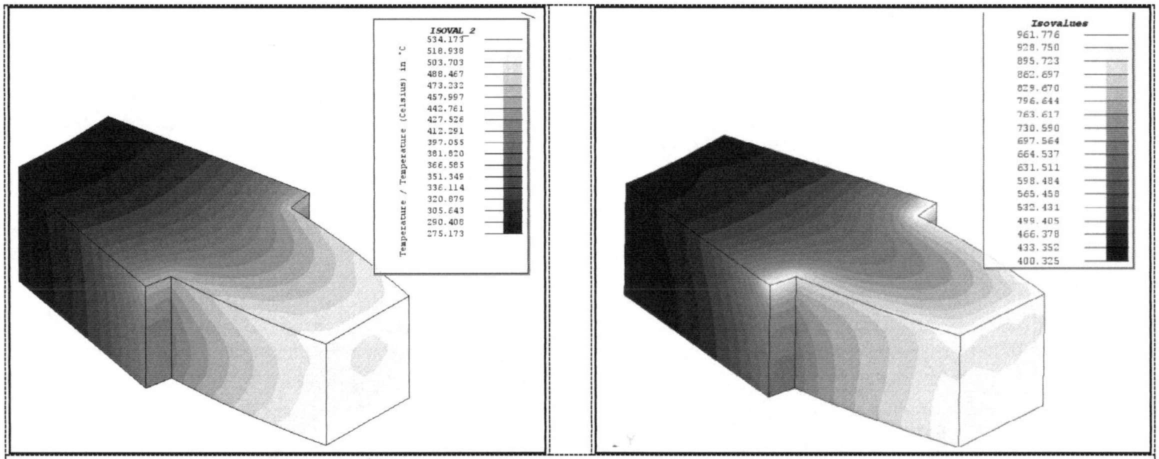

Figure 9. CDFIH process. Temperature distribution: MF heating (left) and HF heating (right).

And the second and the third case the SFIH process were analyzed for the same current and frequency (I_{MF} = 1385 A, f_{MF} = 36 kHz; I_{HF} = 500 A, f_{HF} = 240 kHz) and for the longer time of heating t_{MF} = 6.6 s, t_{HF} = 1.5 s. Temperature distribution after SFIH heating was presented in left side (MF) and right side of Fig. 10.

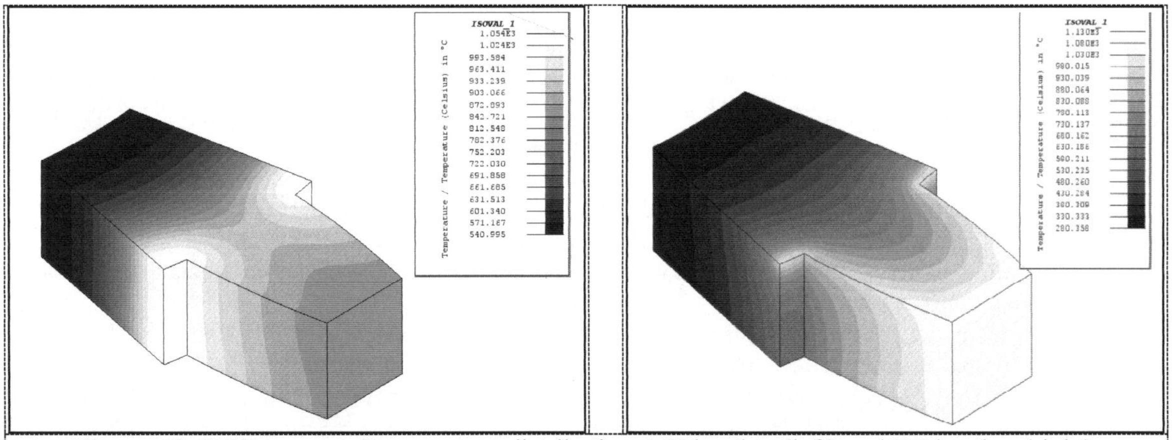

Figure 10. SFIH process. Temperature distribution MF heating (left) and HF heating (right).

For t_{MF} = 6.6 s the average temperature within the tooth is $T_{av} = 935^0$C. Taking into account calculated velocity of heating v_{ih} = 140 K/s the final temperature was sufficiently higher than Ac_{3m}. But the non-uniform temperature distribution was characteristic. At the top of the tooth the average temperature is $T_{avtop} = 820^0$C only. For t_{HF} = 1.5 s the average temperature in the thin hardened zone along the working surface of the tooth is $T_{av} = 953^0$C. Taking into account that for the case calculated velocity of heating v_{ih} = 620 K/s the final temperature was still higher than Ac_{3m}. The temperature distribution in the hardened zone was more uniform in comparison with the previous case. Based upon results (Fig. 9 – 10) hardness distribution was calculated. Computations were compared with the measurements and quite reasonable accordance between them was noticed.

4. Summary

The paper presents the SFIH and CDFIH methods applied for induction hardening of small gear wheels made of steel 41Cr4. Computations were compared with measurements of hardness distribution and the reasonable accordance was achieved, however the expected hardness distribution characterized by thin fully hardened surface zone was obtained for the CDFIH method only. Next research in the area will be aimed on improvement of numerical model in order to shorten time of computations.

5. Acknowledgment

The paper was granted by the National Centre for Research and Development of Poland under the project PBS2/A5/41/2014.

6. References

[1] Rudnev V, Totten G, 2014, Induction Heating and Heat Treatment, *ASM International.* 4C
[2] Rudnev V, 2005, Metallurgical fine points of induction hardening part I, *IndustrialHeatingcom,* pp 1 – 5
[3] Schlesselmann D, Nikanorov A, Nacke B, Galunin S, Schon Z, 2013, Numerical calculation and comparison of temperature profile and martensite microstructures in induction surface hardening processes, *Proceedings of the International Symposium Heating by Electromagnetic Sources*, pp 127 – 134
[4] Barglik J, 2016, Mathematical modelling of induction surface hardening, „*Compel-the international journal for computation in mathematics in electrical and electronic engineering*", Vol. 35, Issue 35, pp 1403 – 1417
[5] Barglik J, Smalcerz A, 2017, Influence of the magnetic permeability on modeling of induction surface hardening, *Compel-the international journal for computation in mathematics in electrical and electronic engineering*", Vol. 36, Issue. 2, pp 555 – 564
[6] Barglik J, Czerwiński M, Hering M, Wesołowski M, 2008, Radiation in Modelling of Induction Heating Systems, *IOS Press*, Amsterdam, Berlin, Oxford, Tokyo, Washington DC, pp 202 – 211

IOP Conf. Series: Materials Science and Engineering **355** (2018) 012016 doi:10.1088/1757-899X/355/1/012016

Self-adaptive multimethod optimization applied to a tailored heating forging process

M Baldan[1], T Steinberg[1] and E Baake[1]

[1]Institute of Electrotecnology, Leibniz University Hanover, Wihlelm Busch Straße 4, 30167 Hanover - Germany

baldan@etp.uni-hannover.de

Abstract. The presented paper describes an innovative self-adaptive multi-objective optimization code. Investigation goals concern proving the superiority of this code compared to NGSA-II and applying it to an inductor's design case study addressed to a "tailored" heating forging application. The choice of the frequency and the heating time are followed by the determination of the turns number and their positions. Finally, a straightforward optimization is performed in order to minimize energy consumption using "optimal control".

1. Introduction

Evolutionary algorithms have been successfully applied in the last few decades for solving optimization and design problems of electromagnetic devices [1]. The goal is to determine values for model parameters that provide the best trade-off in the case of multiple conflicting objects. The combination of the multi-objective optimization procedures with problem-oriented mathematical models allows the development of an effective optimization strategy which can be applied to engineering cases. Since the evaluation of the goal functions requires frequently a FEM model, whose computational cost could be significant, it appears to be crucial to reduce the number of evaluations.

Currently, the non-dominated sorted genetic algorithm (NGSA-II) [2] has received the most attention because of its simplicity and demonstrated superiority over other methods and it has been widely used in the design of electromagnetic devices [3]. Nevertheless, NGSA suffers of some drawbacks when the population and number of iteration must be limited to reduce the time of computation: it can occur, e.g. when the objective functions are evaluated using a numerical method like finite element. Moreover, when the number of individuals in the initial population is limited, the solution sometimes is affected by the choice of the initial population [10]. Particle Swarm Optimization (PSO) [4] is a heuristic search technique that simulates the movement of a flock of birds which aim to find food. The relative simplicity of PSO and the fact that it is a population based technique have made it a natural candidate to be used in design problems [5]. On the other hand, in computational electromagnetics, applications and also modifications of Biogeography-Based Optimization (BBO) are an emerging new field of research [6]. However, it has been proved that it is impossible to develop a single algorithm that is always efficient for a diverse set of optimization problems [7].

Inspired by the work of Vrugt et al. [8], we present an optimization procedure which combines the concepts of simultaneous multimethod search and self-adaptive offspring creation. In the proposed code, contrary to NGSA, PSO and BBO, several mechanisms are together combined for generating offspring. This feature clarifies the meaning of "multi-method" search. The "self-adaptively" comes

Content from this work may be used under the terms of the Creative Commons Attribution 3.0 licence. Any further distribution of this work must maintain attribution to the author(s) and the title of the work, journal citation and DOI.
Published under licence by IOP Publishing Ltd

from the fact that a certain mechanism is preferred if it's more efficient in generating offspring. We intend to apply this innovative optimum search not only to few benchmark problems but also to a real inductor's design case addressed to tailored heating applications. Tailored heating aims at creating a desired temperature profile inside of the work-piece in order to gain desired material properties and/or to minimize energy consumption of e.g. a forging process. Multi-objective optimization aims to find the best combination of geometrical and electromagnetic parameters to fulfil the goals mentioned before.

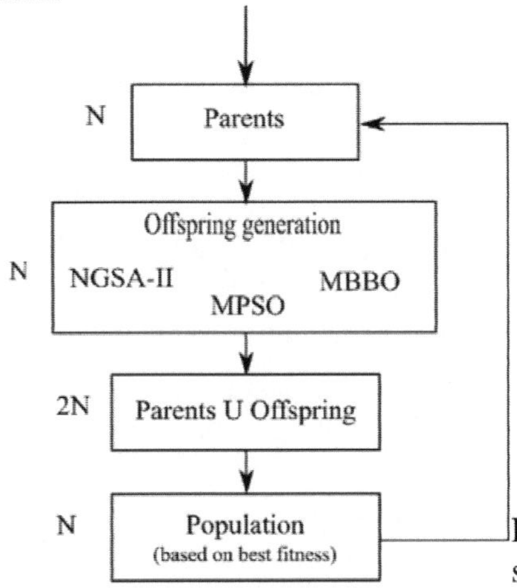

Figure 1. Flowchart of the proposed algorithm. The main structure is the same of the NGSA-II code.

2. Proposed optimization method

We consider a multi-objective optimization problem, with n decision variables: $x = (x_1,..., x_n)$ and m objectives: $f(x) = (f_1(x),..., f_2(x))$ subject to $g_i(x) \leq 0$ $i = 1,2,...,r$ and $h_i(x) = 0$ $i = 1,2,...s$. The presence of multiple objectives give rise to a set of Pareto-optimal solutions, instead of a single solution. A Pareto-optimal solution is one in which one objective cannot be further improved without causing a simultaneous degradation in at least one other objective.

The algorithm here used is a modification of the algorithm AMALGAM proposed by Vrugt et al. [8]. We'll call it, for the sake of simplicity, AMALGAM*. In Figure 1 is shown the main procedure of the optimization code. The algorithm is initialized by using a random population P_0 of size N. Then, to each parent is assigned a fitness value based on the non-dominated sorting mode [2]. The population of offspring $Q_0 = \{Q_0^1,..., Q_0^k\}$, is implemented using k individual algorithms instead of using a single operator. The chance that one offspring is generated by the *k-th* algorithm is proportional to the "specific probability" *k-th* of the algorithm itself. If at the beginning each algorithm has the same "specific probability", at each generation, the values are updated according to the success that algorithms have in producing offspring. The idea of using the "specific probability" is an original contribution of the authors. After offspring's creation, a combined population $R_0 = P_0 \cup Q_0$ of size *2N* is created and ranked using the non-dominated sorting mode. The N members for the next population are chosen according to the rank and the crowding distance of the *2N* elements of R_0. The new population P_1 is then used to create offspring using the already described procedure. This step algorithm is repeated until convergence is reached.

If the vector $S_t^1,..., S_t^k$ indicates the "specific probability" at t generation, $\{N_{t-1}^1,..., N_{t-1}^k\}$ is the number of offspring that, at the previous generation, each algorithm generated and, $\{P_t^1,..., P_t^k\}$ stands for the number of offspring, produced by the *k-th* algorithm, which appear in the current population, the "specific probability" has been calculated as follows (1):

$$S_i^t = \left(P_t^i \Big/ N_{t-1}^i\right)\left(\sum_{j=1}^{k} P_t^j \Big/ N_{t-1}^j\right)^{-1} \tag{1}$$

In order to avoid inactivating algorithms, the minimum "specific probability" value was set to 0.1. In AMALGAM*, we adopted three different algorithms for generation offspring: Non dominated sorting genetic algorithm NGSA-II [2], Multi-objective particle swarm optimization MPSO [4], Multi-objective biogeography-based optimization MBBO [6,9].

NGSA-II uses simulated binary crossover (SBX) and polynomial mutation to create offspring. In the case of MPSO, let $x_h(t)$ denote the position of the *h-th* particle at the iteration *t*, the position of the particle is changed by adding a velocity $v_h(t)$ to the current position (2):

$$x_h(t) = x_h(t-1) + v_h(t) \tag{2}$$

The velocity vector reflects the socially exchanged information. In the case of MPSO, since there is not – like in the single objective case – a best solution, the "best" particle is a random one extracted from an external archive which contains the best $\left|\exp^{2m}\right|$ non-dominated particles (particles are sorted by rank and crowding distance). The best particle's position ever is instead updated when the particle is dominated of if both are incomparable (they are both non-dominated with respect to each other). In MBBO each solution considered is treated as habitat. The offspring are generated by means of two stochastic operators: migration and mutation. Migration is supposed to improve "bad" habitats (which have bad fitness) by sharing features from "good" habitats. Mutation modifies some randomly selected habitats in view of a better exploration of the design space.

In case of constrained optimization, if an offspring (particle) violates one or more constraints in the design space, objective functions are not evaluated. What it is calculated is the "violation" (3):

$$vio_h = \sum_{q=1}^{r} \max(0, g_q(x)) + \sum_{q=1}^{s} \left|h_q(x)\right| \tag{3}$$

Each feasible solution has a better fitness than every other unfeasible solution. Unfeasible solutions are ranked on the basis of their vio_h: the smaller vio_h, the better the fitness.

3. Numerical test

In this section we intend to test AMALGAM* applying it to four analytical cases. We adopt the metrics used in [10] to quantify the performance of the proposed algorithm. We evaluate the error between the approximated and exact one, both in the Pareto front (objective function space) and its correspondent inverse image (variable space). Considering N_{fr} individuals with the lowest rank final front, the geometric distance between each *h-th* point $x = (x_1, \ldots, x_n)$ and its inverse image is (4):

$$d_{x,h} = \sqrt{\sum_{i=2}^{n} (x_{i,h} - f_i^{-1}(x_{1,h}, \ldots, x_{n,h}))^2} \tag{4}$$

Where f_i^{-1} is the *i-th* component of the inverse image of the Pareto front. While, in order to evaluate the error in the objective space, in the case of two objective functions (5):

$$d_{f,h} = \left|f_{2,k} - f_{2,k}(f_{1,k})\right| \tag{5}$$

With $f_{2,k}$ is given the exact expression of the Pareto front. The rms error value of the N_{fr} distances is evaluated in order to compare the performances.

The so called "relative" crowding distance has been also calculated for both the objective and variable spaces. The average "relative" distance of N_{fr} individuals in the variable space is (6):

$$\sigma_x = \frac{1}{N_{fr}-1}\sqrt{\sum_{i=1}^{n}\left[\frac{\max(x_i)-\min(x_i)}{\max(x_i)-\min(x_i)}\right]^2} = \frac{\sqrt{n}}{N_{fr}-1} \quad (6)$$

An analogous expression is used in the objective space, where n is substituted with m (the space's dimension). The average relative distance is called here σ_f. In the variables space, the relative distance between the *h-th* and the *j-th* individuals, both in the front, is (7):

$$d_x(h,j) = \sqrt{\sum_{i=1}^{n}\left[\frac{x_{h,i}-x_{j,i}}{\max(x_i)-\min(x_i)}\right]^2} \qquad h \neq j \quad (7)$$

Again, for the objective space, it's just sufficient to substitute n with m and x with f. The crowding of the front in terms of design variables $cw_x(h)$ is the number of individuals, *j-th*, which are closer to the individual *h-th* than the threshold σ_x (8):

$$cw_x(h) = \sum_{j=1}^{N_{fr}} U_{h,j} \text{ with } U_{h,j} = \begin{cases} 1 & if \ \ d_x(h,j) < \sigma_x \\ 0 & if \ \ d_x(h,j) \geq \sigma_x \end{cases} \quad (8)$$

Also here there is a correspondent definition for the objective space.

3.1. Test functions

We considered four unconstrained test functions [2,10,11] which are summarized in Table 1.

Table 1. Test functions: T1 (A), F2 (B), ZDT1 (C), ZDT2 (D).

	n	Variable bounds	Objective functions	Optimal solutions	Pareto front
A	2	[0,2]	$f_1 = x_1^2 + x_2^2$ $f_2 = (x_1-1)^2 + (x_2-1)^2$	$x_1 = x_2$	$f_2 = f_1 - 2\sqrt{2f_1} + 2$
B	3	$x_1 \in [0,1]$ $x_2, x_3 \in [-1,1]$	$f_1 = x_1 + \frac{2}{3}[x_3 - \sin(6\pi x_1 + \pi)]^2$ $f_2 = 1 - \sqrt{x_1} + \left[x_2 - \sin\left(6\pi x_1 + \frac{2}{3}\pi\right)\right]^2$	$x_i = \sin\left(6\pi x_1 + \frac{i\pi}{n}\right)$ $i = 2,..,n$	$f_2 = 1 - \sqrt{f_1}$
C	30	[0,1]	$f_1 = x_1$ $f_2 = g(x)[1 - \sqrt{x_1/g(x)}]$ $g(x) = 1 + 9(\sum_{i=2}^{n} x_i)/(n-1)$	$x_i = 0$ $i = 2,..,n$	$f_2 = 1 - \sqrt{f_1}$
D	10	$x_1 \in [0,1]$ $x_i \in [-5,5]$, $i \neq 1$	$f_1 = x_1$ $f_2 = g(x)[1 - \sqrt{x_1/g(x)}]$ $g(x) = 1 + 10(n-1)$ $+\sum_{i=2}^{n}[x_i^2 - 10\cos(4\pi x_i)]$	$x_i = 0$ $i = 2,..,n$	$f_2 = 1 - \sqrt{f_1}$

Tests were performed with 20 different starting populations. Here we present the average (over the 20 cases) rms error in both variable and objective spaces, as the most representative indicator in the ability to identify the Pareto front (Table 2). With "Pop" and "Gen" we refer respectively to population's individuals number (called it N before) and to the generations' number. In other words, we decide a priori the number of calls of the objective functions.

Table 2. Comparison of the rms error in the variable and objective space.

Test	Pop	Gen	Code	$err_{rms}(x)$	$err_{rms}(f)$
A	20	20	NGSA-II	0.205	0.068
			AMALGAM*	0.0819	0.0202

IOP Conf. Series: Materials Science and Engineering **355** (2018) 012016 doi:10.1088/1757-899X/355/1/012016

A	20	50	NGSA-II	0.125	0.0333
			AMALGAM*	0.0497	0.0118
B	20	50	NGSA-II	0.478	0.332
			AMALGAM*	0.186	0.0511
B	20	100	NGSA-II	0.408	0.3086
			AMALGAM*	0.1377	0.0285
C	100	200	NGSA-II	0.914	0.698
			AMALGAM*	0.0633	0.0398
D	100	300	NGSA-II	**	**
			AMALGAM*	0.4982	0.1838

In every case AMALGAM* shows an evident superiority compared to NGSA-II. In the case D (**), at every attempt, the code NGSA-II failed the convergence to the real Pareto front. We show here more in details the results in case of test A (Figure 2). A better ability of exploiting the Pareto front is shown by AMALGAM* compared to NGSA-II. Especially in the left side, results obtained with NGSA-II don't cover the overall extension of the Pareto front. This fact is visible in the calculation of the crowding distance: particles amass much more in the case of NGSA-II (Figure 2).

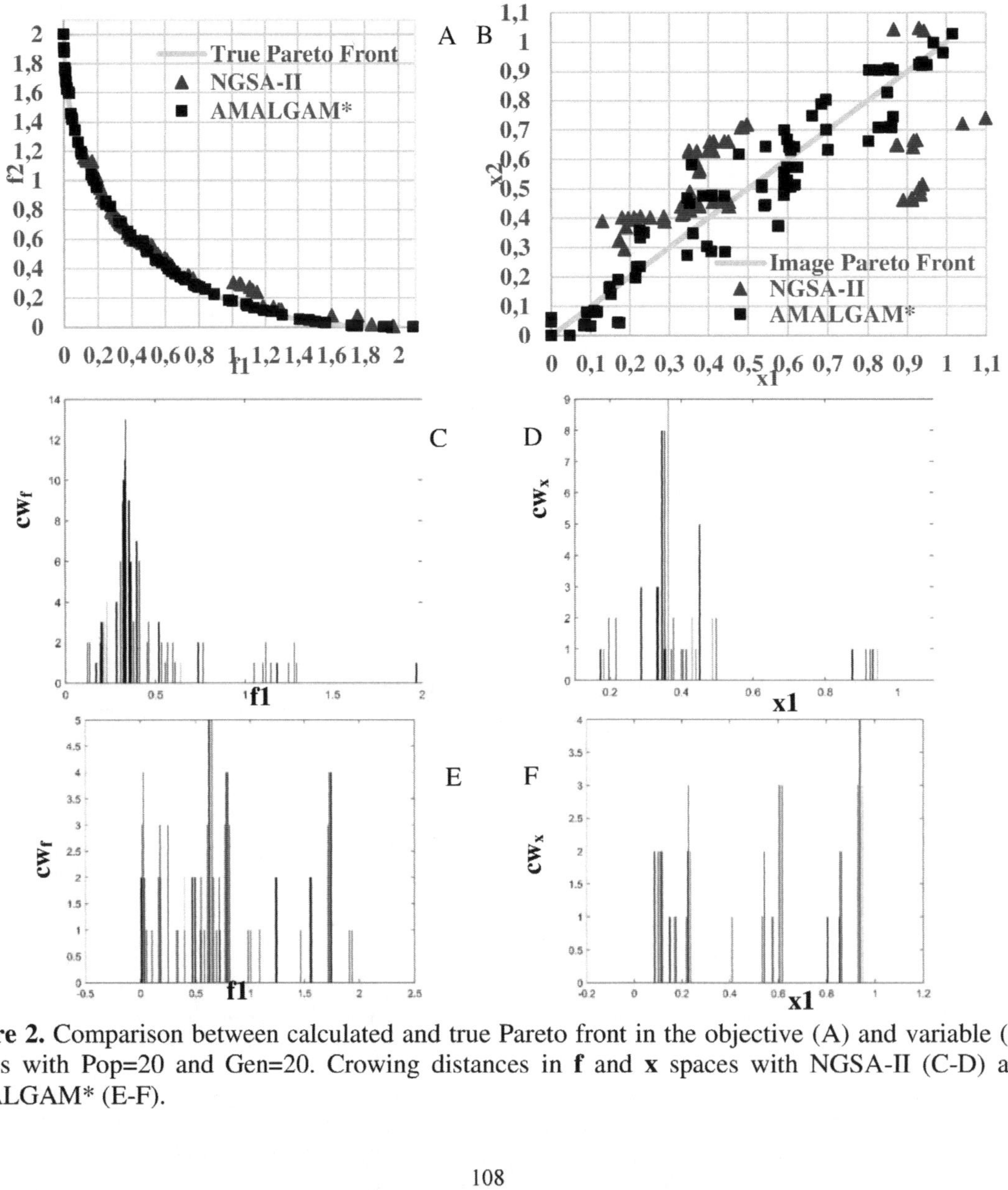

Figure 2. Comparison between calculated and true Pareto front in the objective (A) and variable (B) spaces with Pop=20 and Gen=20. Crowing distances in **f** and **x** spaces with NGSA-II (C-D) and AMALGAM* (E-F).

4. Tailored heating

In literature the concept of “Tailored heating” refers to an innovative heating approach which is addressed to forging processes [12]. Forging a billet from an initially “simple” geometry to a part of “complex” geometry may require several preform stages. In each stage, a pair of dies must be designed and manufactured. Subsequently, a series of mounting and dismounting die-inserts must be performed in the preform stage. This is often time consuming and costly. Consequently, the die forging of parts with complex geometry is only cost effective for large-batch production. Guo et al. [13] and Kayatürk et al. [14] proposed a thermal differential flat-die forging process, in which the parts of the forging billets which are supposed to have a high material flow are heated up to hot forging temperature. Parts which are supposed not to have such high material flow are at “colder” temperatures like 800-900 °C. This contains the basic idea of “Tailored heating”.

The work-piece under analysis is a 42CrMo4 carbon steel billet with 474 mm length and 30 mm diameter. The problem aims to heat up the billet with a desired temperature profile: two “hot” zones with a uniform temperature of 1200 °C alternate with three “cold” zones which are supposed to reach a uniform temperature of 900 °C. It was considered a coupled electromagnetic-thermal model for the field analysis with the package ANSYS®. The design of an inductor which is able to provide the afore described temperature profile is the topic of this work. It represents a challenging task in which AMALGAM* has been applied.

4.1. Choice of frequency and heating time

We want to understand which frequency- heating time couple best fits to our purpose. In [12] good results have been reached using a frequency of 4 kHz and a heating time of 70 seconds. We consider for simplicity a 1D model (Figure 3), in which the inductor is a massive piece of copper and the work-piece a 15 mm wide 42CrMo4 body.

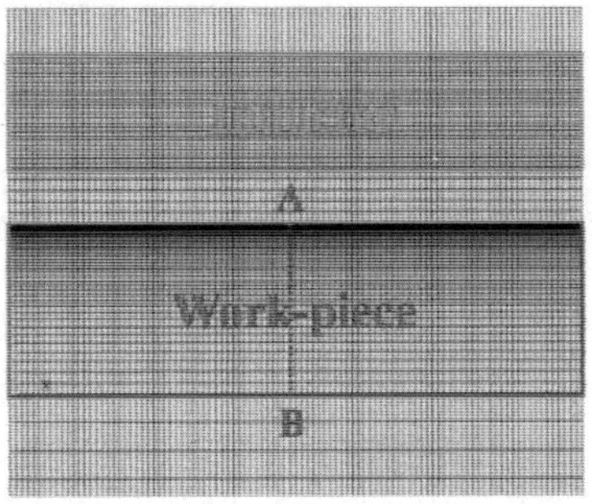

Figure 3. 1D model’s geometry.

Our intention is to obtain a uniform temperature of 1200 °C. Two objective functions are evaluated: the maximum deviation, at the end of the heating, along the line A-B (see Figure 3) from 1200 °C and the energy consumption (9). A fixed current is applied for the entire duration of the heating.

$$f_1 = \max\left|T(l_{AB}, t_{heat}) - 1200^\circ\right|$$
$$f_2 = \int_{t_{heat}} P dt \qquad (9)$$

Three design parameters are under investigation: heating time (t), frequency (f) and current (I) (Table 3). Values of current don’t have a real physical meaning because we are referring to a 1D model.

Table 3. Parameters to be optimized.

t (s)	f (kHz)	I (kA)
40 - 70	4 - 7	4 - 7

Figure 4 summarizes the results of optimization after 100 generations with a population of 20 individuals. While AMALGAM* shows a greater ability in exploring the Pareto front compared to the results obtained with NGSA-II, the minimum value of f1 is practically the same (see Table 4).

Table 4. Individual with the minimum value of f_1 (best heating accuracy).

	t (s)	f (kHz)	I (A)	f_1 (°C)	f_2 (Wh)
NGSA-II	69.8	4	4634	50.9	1682
AMALGAM*	70	4	4626	50.0	1681

As expectable, the maximum homogenization is obtained with longest time and the smallest frequency. What is interesting to see, is that, moving along the Pareto front, with a frequency of 4 kHz and a heating time of 63 s, f1 values 55°C, which is only 5°C more than the found optimum with minimum f1. On the other hand, the energy consumption is in this case 1655 Wh (1.6% less) and the time is drastically reduced. For this reason, we adopt a frequency of 4 kHz and a heating time of 63 s. In Figure 5 instead are shown the trends of the "specific probability" during the 100 generations of the three optimization algorithms used in AMALGAM*. If at first NGSA-II and MBBO are dominant, the Particle Swarm shows a better ability in exploiting the Pareto front. That's why it has the maximum specific probability in the last generations.

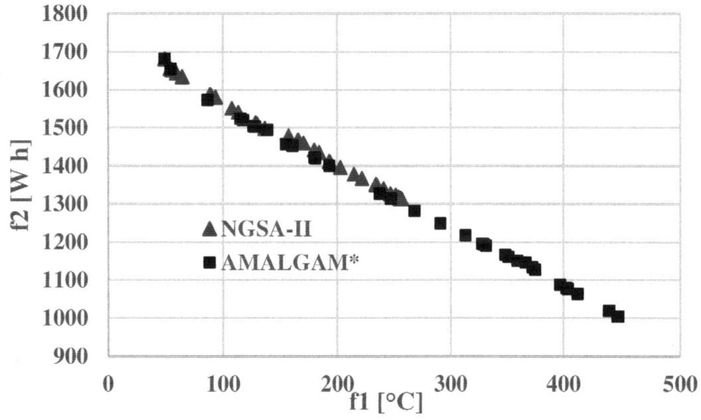

Figure 4. Approximated Pareto front after 100 generations. NGSA-II's Pareto front is much more limited.

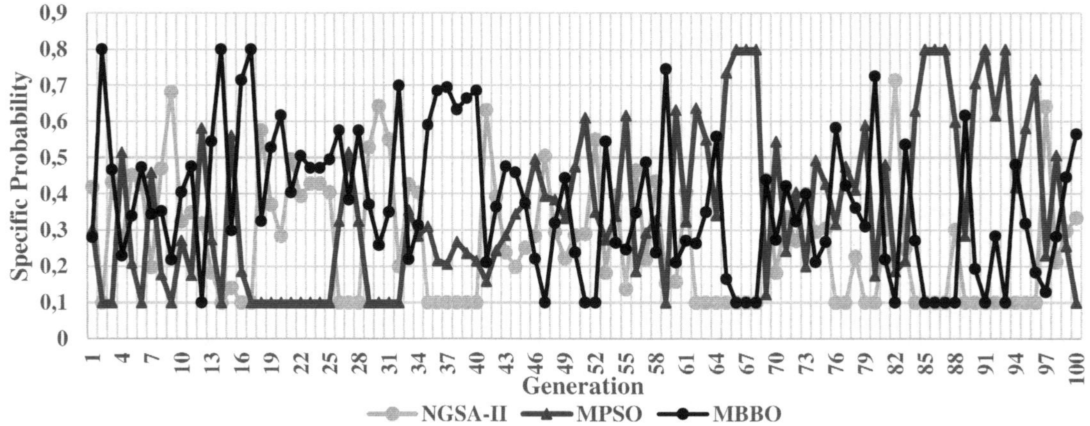

Figure 5. Specific probability of NGSA-II, MPSO and MBBO during the 100 generations.

4.2. Inductor's design

The next step is the design of the inductor. Since it's very difficult to define a priori the inductor's configuration (number of turns, air gaps between them…), in the model proposed the inductor is substituted by five nonconductive areas in each of them a uniform current density is applied. In order to speed up the solution, the numerical model consists of one fourth of the system (Figure 6). As design variables, the length l_i and the current density j_i applied to each area are considered. l_0 indicates the position of the first area from the middle of the billet. In total 11 variables are under analysis (Table 5).

Table 5. Design variables in the inductor's design problem.

l_0 (mm)	l_1	l_2	l_3	l_4	l_5	j_1 [A/m^2]	j_2	j_3	j_4	j_5
1-10	20-60	5-20	70-150	5-40	40-150	0.5-1.0 10^7	0.1-0.8 10^7	0.8-1.35 10^7	0.1-0.9 10^7	0.6-1.3 10^7

Figure 6. Geometry of the model.

The aim is to obtain a uniform temperature of 1200 °C in the hot zone and 900 °C in the cold ones. As visible in Figure 6, we don't take into account the temperature between different zones, in a transition zone 15 mm long. Three objective functions are evaluated (10):

$$f_1 = \sum_{i=1}^{W} \left| T(x_i, y_i, t_{heat}) - 900° \right|$$
$$f_2 = \sum_{i=1}^{Q} \left| T(x_i, y_i, t_{heat}) - 1200° \right|$$
$$f_3 = \sum_{i=1}^{Z} \left| T(x_i, y_i, t_{heat}) - 900° \right| \qquad (10)$$

Where W and Z is the number of nodes respectively in the cold zone at the "centre" and at the "side" of the work-piece. Q is the number of nodes in the hot zone. Each objective function is the sum of the differences, estimated in each node, between the desired and calculated temperature at the end of the heating. Although this choice doesn't impose any limit to the maximum deviation between final and desired temperature which can be reached in a certain zone, it seems to be more effective in the event that the temperature distribution is only coarsely known. The huge number of variables require a great amount of goal functions' evaluations. A population of 40 individuals after 70 generations has given these results in terms of goal functions (Figure 7).

Table 6. Best solution for the inductor's design.

l_0 (mm)	l_1	l_2	l_3	l_4	l_5	f_1 (°C)	f_2	f_3
2.4	20.1	12.9	114.0	18.4	109.5	25.45 10^3	65.17 10^3	11.45 10^3
	j_1 (A/m^2)	j_2	j_3	j_4	j_5			
	0.999 10^7	0.129 10^7	1.118 10^7	0.433 10^7	0.798 10^7			

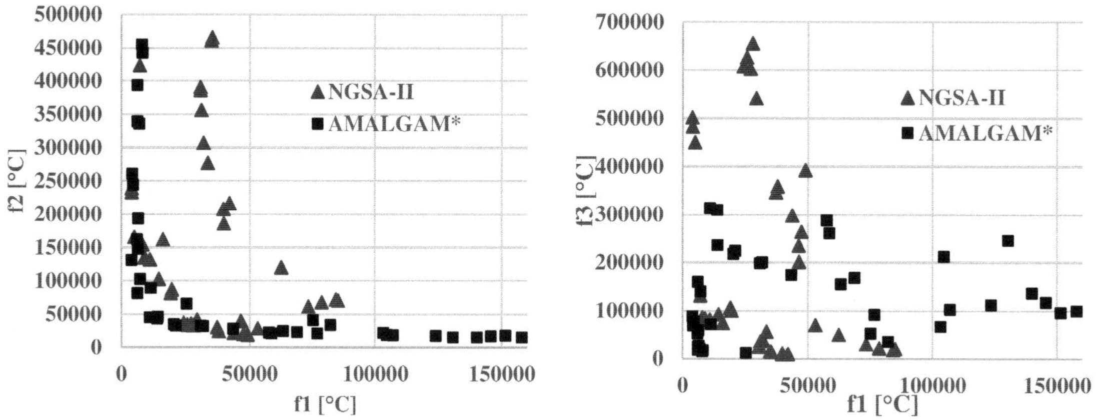

Figure 7. Results of optimization after 2800 goal functions evaluations.

For the design of the inductor we consider the closest solution to the "utopia" point (Table 6). From the technological point of view f_1, f_2 and f_3 are not fully representative whether the solution is suitable or not. Substantially f_2 with a sum of 65.17 10^3 keeps a potential to be minimized and thus become

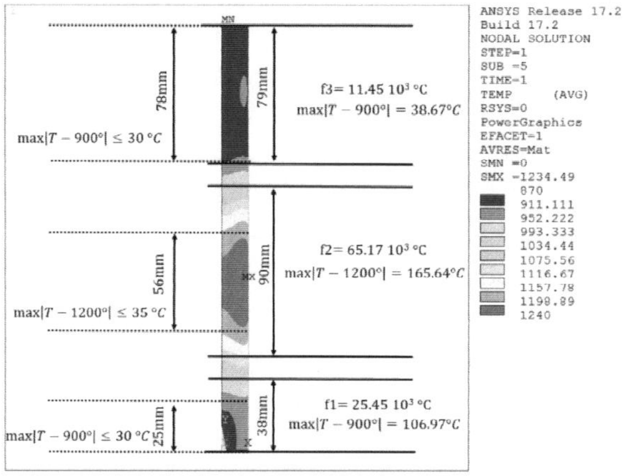

Figure 8. Final temperature profile relative to the best solution.

more closely to an optimum solution. In Figure 8 is shown the final temperature distribution relative to the best design point.

Especially in the "hot" zone, the temperature in the whole area is far from being uniform. A quite flat temperature field has been reached only in the middle part of the "hot" zone, instead the maximum deviation from 1200°C is achieved at the edges. In Table 7 are summarized both the maximum deviations from the goal temperature and the largeness of zone in which the temperature difference doesn't exceed 35°C. Referring to the "hot" zone, over a length of 56 mm the temperature doesn't differ more than 35° from 1200°C.

Table 7. Best solution for the inductor's design.

	Central "cold" zone	"Hot" zone	External "cold" zone
Length (mm)	38	90	79
Max Deviation (°C)	106.97	165.64	38.67
Length (mm)	25	56	78
Max Deviation (°C)	< 30	< 35	< 30

4.3. Real Inductor

As already done in [12], we need to “translate” this configuration in a real inductor. Each area could be seen as one turn coil. For designing the real inductor, assuming a series connection between turns, it’s possible to convert every one-turn coil in an equivalent multi-turns coil. The total equivalent current (current density times area) which flows in each area is visible in Table 8.

Table 8. Total equivalent current.

I1 (A)	I2 (A)	I3 (A)	I4 (A)	I5 (A)
2008	166	12745	797	8738

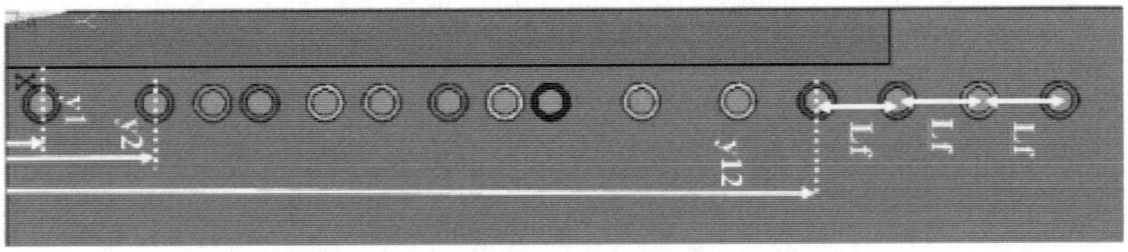

Figure 9. Geometry of a generic inductor.

The number of turns in each zone has been chosen as follows: 2 in the first, 7 in the second and 6 in third. In total there are 15 turns in a half inductor, which are equally distributed in each area. The resulting temperature profile is bell-shaped at the centre of the “hot” zone due to the fact that there the magnetic field is maximum. For this reason, a new sub-optimization is performed, in which, the positions of the turns are modified in order to get a more satisfying temperature profile. In this case 13 variables are under analysis: the y positions of the first 12 turns plus the (assumed) constant distance of the last three turns (see Figure 9).

Since the overlapping between turns is not allowed, we’re looking at a constrained optimization case. We set a minimum distance between turn and turn of 1 mm. Using a coil with an external diameter of 10 mm, the following inequalities must be true (11):

$$y_{i+1} - y_i \geq 11 \text{ mm, with i=1,...,11} \tag{11}$$

In Table 9 the variables’ domains and the solution which best meets our purposes are summarized. Goal functions are the same that have already been used (10). A population of 30 individuals had been run for 100 generations. The total number of evaluations is 2384 for the NGSA-II and 2375 in the case of AMALGAM*.

Table 9. Design parameters and best solution obtained.

	y_1	y_2	y_3	y_4	y_5	y_6	y_7	y_8	y_9	y_{10}	y_{11}	y_{12}	L_f
Min	8.5	38	50	64	80	95	110	127	142	165	190	210	18
Max	12	43	60	74	90	105	125	140	155	180	201	225	26
Best	9.5	40	55.4	68	85.5	100.7	118.5	133.5	144.5	169.8	195.7	217.1	22.0

IOP Conf. Series: Materials Science and Engineering **355** (2018) 012016 doi:10.1088/1757-899X/355/1/012016

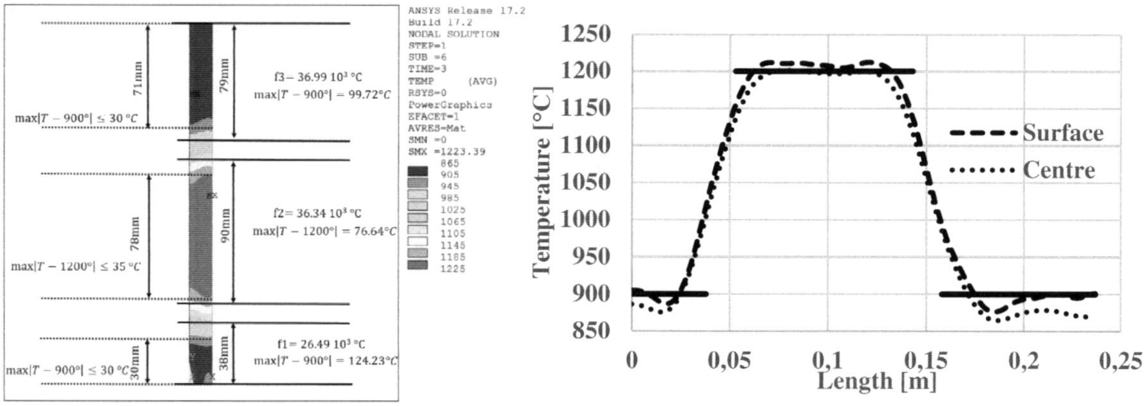

Figure 10. Temperature field (left) at the end of the heating. Temperature profile (right) along the length at the surface and in the centre.

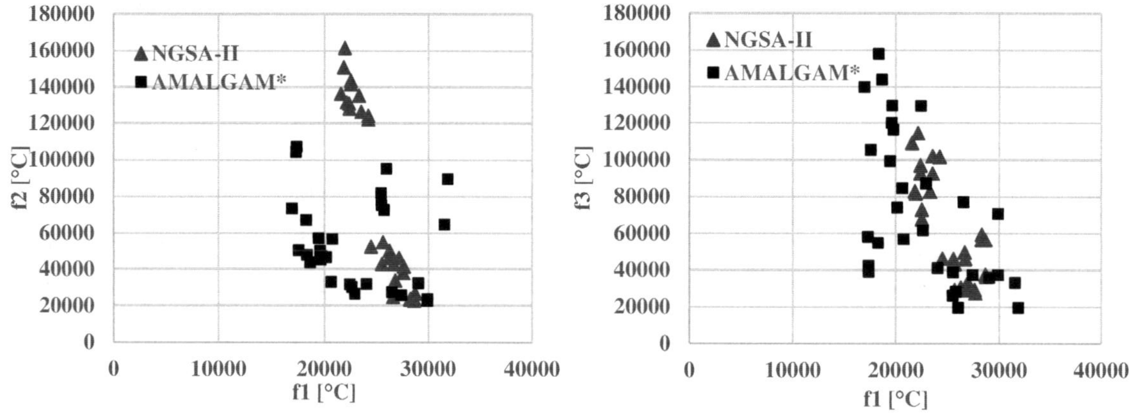

Figure 11. Results of optimization in the real inductor case.

Table 10. Best solution for the inductor's design.

	Central "cold" zone	"Hot" zone	External "cold" zone
Length (mm)	38	90	79
Max Deviation (°C)	124.23	76.64	99.72
Length (mm)	30	78	71
Max Deviation (°C)	< 30	< 35	< 30

Figure 10 shows the temperature distribution of what is supposed to be the best result from the optimization procedure. A solution with relative small f2 value is here favourite (Figure 11). Nevertheless, even in the best solution the temperature field diverts considerably from the wished one at the edges of each zone. Table 10 sums up maximum deviations and describes the zones extensions which have a maximum temperature deviation below 35°C. Values are remarkably greater to those in Table 7, which justifies the necessity to perform a sub-optimization. Actually, it has not been found a temperature field in which the "transitions zones" are only 15 mm wide. Transitions in our case study take values of approx. 29 mm, to whose correspond a gradient of 100 °C/cm.

4.4. Optimal control

In this last section we intend to apply the concept of optimal control to the already designed inductor. In general, the ability to control the heating process means a capability to influence a temperature field in order to achieve a desired goal. The temperature distribution can be affected trough a set on control inputs. By choosing in every instant the value of each control input, one can modify the temperature distribution and dynamic behaviour of an induction heating system [15]. To choose certain control properly, one needs to determine a cost function, i.e. a function reflecting technical and economic efficiency of system. For example, when maximum productivity is required, a minimal total heating time can be considered as cost function [16].

We intend to reduce the total energy consumption applying a two stage process: in the first the heating under maximum power is performed, in the second no power is provided (soaking period). It is possible to prove mathematically that in case of a static billet which has to be uniformly heated up, time-optimal control consists of alternating stages of heating with maximum power and subsequent soaking. Therefore, the shape of optimal control algorithm is known, but the number of stages and their durations remain unknown. The greater is the number of stages, better is the heating accuracy achievable. In the present work we refer to the simplest case with one heating and one soaking interval, even if is not proven that this choice is optimal.

Typically, optimal control applications are performed with a defined voltage. That means the heating is characterized by a first interval where the inductor is fed by the maximum voltage available and by a soaking interval. Since in this work we have always thought in terms of current, also in this section we assume the current being the input parameter. In the inductor's optimization a current of 1230 A was given. Supposing we can provide a maximum current of 1500 A – limit given by the maximum power available – we investigated how the energy consumption varies with the heating accuracy.

In summary, the intervals in which vary the two design variables are (Table 11):

Table 11. Possible intervals for heating and soaking time.

t_{ON} (s)	t_{OFF} (s)
30 - 65	0.1 - 15

Goal functions take into account the heating accuracy (f_A) and the energy consumption (f_B) (13). As in Section 4.2, W, Z and Q are the total number of nodes in the "cold" and "hot" zones.

$$f_A = \sum_{i=1}^{W} \left|T(x_i, y_i, t_{heat}) - 900\,°\right| + \sum_{i=1}^{Q} \left|T(x_i, y_i, t_{heat}) - 1200\,°\right| + \sum_{i=1}^{Z} \left|T(x_i, y_i, t_{heat}) - 900\,°\right|$$

$$f_B = \int_{t_{heat}} P dt \qquad (12)$$

Since an excessive overheating of the billet would be not acceptable, a maximum admissible temperature of 1300 °C was set. If during the heating, any point of the billet overcomes 1300 °C, the solution is considered not feasible. Figure 12 shows the Pareto fronts' trend in the case of both, NGSA-II and AMALGAM*, with a population of 30 individuals, after 50 generations.

The estimated Pareto front obtained with AMALGAM* dominates the NGSA-II's one. The best heating accuracy (minimum f_A) reached by the two codes is shown in Table 12. In the "normal" heating mode described in the paragraph 4.3, f_A would value 100671 °C and f_B 355.8 Wh. If on one side the results of optimal control don't bring any improvement in terms of heating accuracy, on the other hand, with a small worsening of the temperature profile, an important reduction of time and energy consumption could be obtained. The soaking time tends indeed to vanish the temperature gradient between zone and zone.

Table 12. Minimum value of f_A in the two codes.

	t_{ON} (s)	t_{OFF} (s)	f_A (°C)	f_B (Wh)
NGSA-II	42.0	6.6	109342	353.0
AMALGAM*	41.4	5.5	108882	350.2

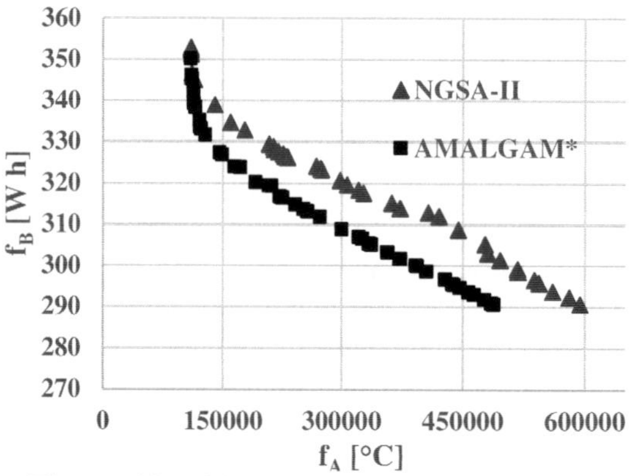

Figure 12. Objective functions after 50 generations.

5. Conclusions

Using a multimethod optimization search has been found to be a more successful solution in both test and real cases compared to the well-known NGSA-II. Combining different offspring generation methods in a self-adaptive way enables, at each generation, to prevail the most effective ones. In AMALGAM*, if at first NGSA-II and MBBO hold a high specific probability, when the Pareto front has been identified MPSO remains the most active method. The proposed algorithm improves Pareto front estimation in problems for which objective functions evaluation are computationally expensive. Especially for the tailored heating process, a new inductor design including an optimized set of parameters has been investigated from one benchmark experiment followed by an extensive parameter study. Here it could be shown, that multimethod optimization can be applied successfully in order to minimize the energy consumption by fulfilling the desired temperature profile.

References

[1] Di Barba P, Dughiero F, Forzan M and Sieni E 2016 Migration corrected NGSA-II for improving multi-ojective design optimization in electromagnetics *Intern. J. of Applied Electromagnetics and Mechanics* **51** 161-72

[2] Deb K 2002 A fast elitist multi-objetive genetic algorithm: NGSA II *IEEE Transaction on evolutionary compuation* **6 N. 2** 182-97

[3] Di Barba P, Forzan M and Sieni E 2014 Multi-objective design of a power inductor: a benchmark problem of inverse induction heating *COMPEL The Intern. J. for Compuation and Mat.in Electrical and Electronic Engineering* **33** 1990-2005

[4] Reyes Sierra M and Coello Coello C A 2006 Multi-objective particle swarm optimization: a survey of the state of art *International J. of Computation Intelligence Research* **3 N. 2** 287-308

[5] Nando J and Yahya R S 2008 Particle swarm optimization for antenna designs in enginnering electromagnetics *J. of Artificial Evolution and Applications* **2008** 22-10

[6] Di Barba P, Dughiero F, Mognaschi M E, Savini A and Wiak S 2016 Biogeography-inspired multi-objective optimization and MEMS design *IEEE Transactions on Magnetics* **52 N.3**

7201504
[7] Wolpert D H and Macreasz W G 1997 No free lunch theorems for optimization *IEEE Transactions on evolutionary computations* **1 N.1** 67-82
[8] Vrugt J Robinson B 2007 Improved evolutionary optimization from genercially adaptive multimethod search *PNAS* **104**
[9] Dan S 2008 Biogeography-based optimization *IEEE Transactions on evolutionary computation* **12 N.6** 702-703
[10] Sieni E Di Barba P and Forzan M 2016 Migration NGSA: method to improve a non-elitist searching of Pareto front with applications in magnetics *Inverse Problem in Science and Engineering* **24 N. 4** 543-66
[11] Li H and Zhang Q 2009 Multiobjective optimization problems with complicated Pareto sets, MOEA/D and NGSA-II *IEEE Transactions on evolutionary computation* **13 N. 2** 284-302
[12] Baldan M Steinberg T and Baake E 2017 Numerical and practical investigation of tailored heating process for forging parts *Proceedings of XVIII UIE Congress* Hannover 338-345
[13] Guo Y F and Hua M and Osam F H 1997 A preliminary analysis of thermal differential forging – a ring hearer or locally heated cylindrical rod *Proceeding of the 24th International Thermal Conductivity Conf. and 12th Intern. Thermal Expansion Symposium* Pittsburgh PA 731-45
[14] Kayatürk K Kurt A Weidig U Steinhoff K Tekkaya A 2001 Simultaneous cold and hot forging in a single forming step – principle, possibilities and limitations *Proceedings of 3th International Conference on Industrial Tools* Slovenia 97-100.
[15] Rapoport E Pleshivtseva Y 2007 Optimal Control of Induction Heating Processes Taylor & Francis
[16] Pleshivtseva Y Zaikina N Nacke B Nikanorov A Time-optimal control of energy-efficient heating of alluminium billets rotating in DC magnetic field

IOP Conf. Series: Materials Science and Engineering **355** (2018) 012017 doi:10.1088/1757-899X/355/1/012017

Modelling of thermal stresses in bearing steel structure generated by electrical current impulses

M Birjukovs[1], A Jakovics[1] and W Holweger[2]

1 Department of Physics, University of Latvia, Zellu Street 23, Riga LV-1002, Latvia
2 Tribology Fundamentals, Schaeffler Technologies GmbH & Co. KG, Industriestraβe 1-3, Herzogenaurach 91074, Germany

E-mail: `michael.birjukov@gmail.com`, `andris.jakovics@lu.lv`

Abstract. This work is the study of one particular candidate for white etching crack (WEC) initiation mechanism in wind turbine gearbox bearings: discharge current impulses flowing through bearing steel with associated thermal stresses and material fatigue. Using data/results from previously published works, the authors develop a series of models that are utilized to simulate these processes under various conditions/local microstructure configurations, as well as to verify the results of the previous numerical studies. Presented models show that the resulting stresses are several orders of magnitude below the fatigue limit/yield strength for the parameters used herein. Results and analysis of models provided by Scepanskis, M. et al. also indicate that certain effects predicted in their previous work resulted from a physically unfounded assumption about material thermodynamic properties and numerical model implementation issues.

1. Introduction

WEC formation is a common yet poorly understood fatigue failure mode in machinery components subject to cyclic and/or transient loads. WEC occurrence is widespread (automotive gearboxes, industrial gearboxes, mill drives, cranes, etc.), but it is especially frequent in wind turbine gearbox bearings. These undergo breakdowns much earlier than predicted by rolling contact fatigue (RCF) models, which implies that other factors are also responsible for shorter bearing lifetimes (6 – 24 months VS the desired 20 years). A solid, rigorous explanation of this phenomenally rapid bearing breakdown in wind turbines eludes the scientific community to this day. The resulting annual expenditures in the wind energy industry due to turbine standstills and maintenance are tremendous (especially in case of relatively distant offshore installations), which is why investigations in this field are of utmost importance. [1]

Most bearings unavoidably fail via spallation damage at the end or their operational lifetimes. One of the standard measures of these is the L_{10} life: the RCF life of a bearing for a given operating condition at which statistically 90 % of bearings survive. The failure type in question here, dubbed the white structure flaking (WSF), typically sets in at 5 – 20 % of expected L_{10} values [1]. WSF is due to microstructural alterations in bearing races at about 1 mm depth beneath the contact surface. The resulting altered structure areas are known as white etching areas (WEAs) and these are usually associated with cracks found in close proximity,WECs, which come in two varieties: surface (originate from contact surface) [2,3] and subsurface (the majority of WECs have no connection to contact surface) [4,5,6]. The reason for such peculiar nomenclature is that the appearance of WECs and WEAs becomes distinctly white colored

Content from this work may be used under the terms of the Creative Commons Attribution 3.0 licence. Any further distribution of this work must maintain attribution to the author(s) and the title of the work, journal citation and DOI.
Published under licence by IOP Publishing Ltd

after the containing steel structure is subjected to nital etching during the post failure structure analysis of cross sectioned bearings [1]. It has been shown that WEA consists of nano ferrite grains and is considerably harder than the surrounding steel matrix [1,7,8], which is in agreement with the Hall-Petch relationship.

Considering that pure RCF cannot be responsible for premature failure, several hypotheses and theories proposing different WEC formation assisting mechanisms have been developed and are clearly outlined in the recent review article by Evans [1]. These include steel corrosion (during standstill periods and due to maintenance contamination), lubricant water contamination [9-11], tribochemical processes (such as hydrogen generation within lubricant film) [12-15], bearing slip (during accelerations/decelerations, intentional or otherwise), vibrations [10,16], electrical currents (due to lightning strikes, lubricant self charging/discharging and other stray currents) [1,17-19], additional bending stresses and impact loads (for instance, due to bearing misalignment, wind gusts, irregularities in electrical grid power supply regime) [10,20,21]. The current objective is to identify the root WEC initiation cause(-s) among the above mentioned factors, but given that these may or may not influence the RCF process simultaneously (degrees of influence depend on operational conditions), it is not at all trivial. It is therefore vital to ascertain the severity of each of the potential WEC initiators and, if possible, eliminate the irrelevant ones from the rather long list of suspects.

A recent investigation by Smelova et al. [22] on an RCF test rig has demonstrated that premature cracks form well before and in absence of material transformation (including WEAs). It must be noted that cracks that precede WEAs were, although under different conditions, previously demonstrated by Gould and Greco [23] and others, but such cracks were preceded by dark etching area (DEA) formation. This is a milestone result, because a long standing topic for debate was whether WEAs are merely symptoms of WECs or the other way around: WECs are caused by stresses localized at WEA/steel matrix boundaries due to significant hardness differences and thus mechanical incompatibility [1]. With this controversy seemingly resolved, one of the avenues of research has been potentially eliminated, which is important because of the sheer number of theories and working hypotheses that are currently in play.

Hydrogen generation and subsequent uptake by steel results in a considerable decrease of material yield strength and, consequently, fatigue limit, as well as other effects [12-15]. Damage to steel surface facilitates hydrogen penetration through microcracks: an effect further enhanced by vibrations [24]. However, it has been reported repeatedly that industrial tests, including those by *Schaeffler Technologies*, have produced WSF in bearings lubricated with virtually hydrogen-free oils; some tests were conducted in vacuum and eliminated water/air contamination [25,26]. These results imply that hydrogen diffusion into steel in not a requisite for premature bearing failure, which raises a question of just how relevant this process actually is, if at all, under certain conditions [1]. Note that the above mentioned rig tests also eliminated additional transient loads usually present during wind turbine operation.

Even more importantly, tests have indicated that WEC formation could be accelerated by applying external voltage to bearings, between inner and outer rings, as seen in publications by Smelova et al. [22], Loos et al. [18] and other cases, with or without external voltage [25,27]. More recently, Scepanskis et al. [28] have reported empirical evidence of voltage self-generation in a lubricated tribocontact. While the exact relationship between electrical currents and WEC formation is unknown, these experimental facts suggest that it could be a major influence and should be regarded seriously.

In contrast to extensive experimental research, very few theoretical or modelling works regarding WECs have been published to date, most of which deal with contact mechanics, material transformations, hydrogen diffusion, crack propagation (not initiation) or attempt to explain how WEAs are formed (for instance, see [16,20,29-39]; while [36] is not directly related to bearing steels, in author's opinion, it demonstrates microstructure modelling capabilities

IOP Conf. Series: Materials Science and Engineering **355** (2018) 012017 doi:10.1088/1757-899X/355/1/012017

very well). This is especially true in case of electricity, in which, to the author's knowledge, no modelling results were published that would showcase the influence of electrical current on fatigue acceleration, except for the article by Scepanskis et al. [17], wherein it has been proposed that electrothermally generated stress is concentrated at defect sites, such as carbides, due to discharge current generated Joule heat where current flows around defects.

That publication displays some very intriguing, but equally alarming results: predicted thermal stresses (on the order of 10^2 MPa) are dangerously close to common bearing steel yields strengths (also on the order of 10^2 MPa) and temperature increase is on the order of tens of kelvins. Significant carbide grain and surrounding matrix deformations are observed and the model in question predicts that they are severe enough to lead to the formation of nanovoids near carbide grains. Voids in vicinity of defects are, in fact, readily found in failed bearing steel microstructure and are obvious crack initiators: for instance, see [17, Fig. 8].

The major problem is that almost no estimates for the microstructure material physical properties are available, the exceptions being somewhat rough values, hence the relative infrequency of theoretical/modelling articles. The authors of the present work believe addressing this issue is one of the priorities, since mircostructural process modelling has the potential to provide many new insights into premature crack initiation drivers and dynamics. There is also little to no data on characteristic electric current densities and discharge durations, so Scepanskis et al. [17] assume certain values, which may have been exaggerated.

Nevertheless, [17] showcases a very important result that opens up another avenue of research. The authors of present work believe that it certainly warrants further investigation and also, being the only result, validation, hence the present publication. Since the work of Scepanskis et al. [17] was published, more rig tests were conducted to investigate WEC promotion by electricity [18,22], notably the work of Loos et al. [18], wherein electrical discharges in bearings are studied: it is possible to derive a rough discharge duration estimate from the data presented therein.

With that in mind, the authors present the development of a model for thermal stress accumulation within the steel structure due to discharge currents known to flow through bearings during operation. The objectives of this study are to construct a more generalized model including a wider range of effects and to verify the previously obtained results. In case of a positive investigation outcome, electrothermal WEC initiation could be confirmed as a key bearing failure acceleration mechanism. This work would then serve as ground zero for further efforts to predict the extent of damage inflicted to bearing steel by stray electric currents.

2. General modelling framework

One must note that currently there is little to no information regarding the range of magnitudes of stray currents in operational gearboxes. There exist hypotheses as to the origins of these discharges, but none have been verified with certainty. However, rig tests where cyclic loads were combined with artificially applied voltages across bearings [18,22] provide grounds for initial attempts to ascertain the magnitude of current induced stresses without delving into the details of current generation/propagation.

Steel microstructure tends to be rather complex, so for generality purposes models presented herein assume a spherical carbide (grain) within the otherwise homogeneous martensite phase steel structure by default, with various model to model modifications (discussed below). A thin relatively heightened thermal/electric resistance interface layer between the grain and the surrounding structure commonly seen in alloys is accounted for. Discharge current is assumed to flow through a straight cylindrical channel containing the carbide grain.

In light of the work by Scepanskis et. al. [17], presented models will match material parameters used therein for verification purposes. Discharge duration (t_{dis})/delay (t_{del}) are 1 μs and 100 μs respectively loosely based on the data obtained by Loos et al. [18], which is used as a reference due to lack of information about the actual discharge times in wind turbine bearings. All presented calculations are based on parameters in table 1.

Table 1. Relevant material properties.

Parameter, units	Martensite steel	Carbide grain	Material boundary
σ, electric conductivity, S/m	$100 \cdot 10^5$	10^5	10^5
k, thermal conductivity, $W/(m \cdot K)$	50	1	0.01
c_p, specific heat capacity, $J/(kg \cdot K)$	400	600	600
α, thermal expansion coefficient, $1/K$	$15.0 \cdot 10^{-6}$	$7.6 \cdot 10^{-6}$	N/A
ρ, density, kg/m^3	7500	7500	7500
E, Young's modulus, Pa	$2.1 \cdot 10^{11}$	$1.9 \cdot 10^{11}$	N/A
ν, Poisson ratio, $none$	0.30	0.32	N/A

Boundary layer thickness is assumed to be 0.01 μm. Note that only rough estimates of the boundary layer parameters are available. Since the model only deals with an isolated defect, the surrounding structure is assumed to be regular in the sense that grains are spaced by separating distances of 10 μm (some of the longest seen in micrographs, as evident from [1] and references therein) in an array of cells where a grain is surrounded by steel matrix, current channel being directed through the cell. Grain radius is set to $r_g = 0.5$ μm, current density is set to 10^8 A/m^2 throughout the discharge period.

3. Proposed models

3.1. Analytical model

Proposed model hierarchy schematic representation is found in figure 1.

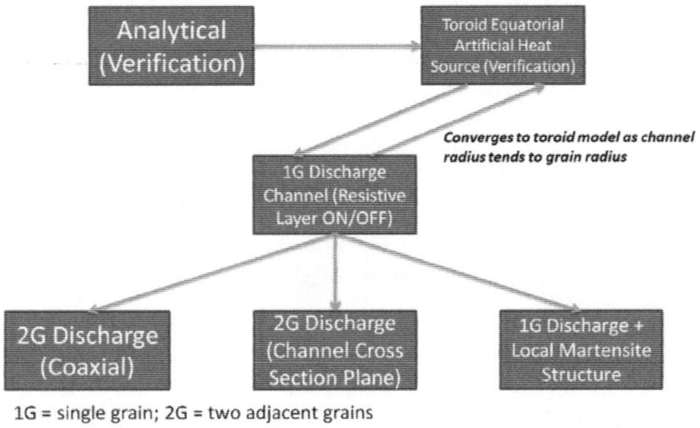

Figure 1. Model hierarchy schematic representation.

First, a simplified analytical model is constructed. Its purpose is twofold: firstly, it provides order of magnitude estimation and approximation of temperature distribution profiles and accumulated thermal stresses (time evolution) within the domain of interest to later verify numerical model results; secondly, it assists in numerical model definition and establishment of validity bounds.

The assumption is that when current flows around a spherical defect, Joule heat is concentrated near the grain/steel boundary along grain equator. Current density magnitude at the equator is estimated from the discharge channel (radius r_c) base/grain restricted cross section areas ratio for r_c/r_g close to 1. Due to rotational symmetry with respect to current channel axis, temperature profile time evolution is solved for using a one dimensional radial line (coplanar to heat source) approximation, wherein two domains are considered: grain and steel, where a point-like Joule heat source is placed at the interface. To further simplify the problem, it is solved separately for each domain on a semi-open interval, which is essentially equivalent to placing a layer with zero thermal conductivity between the materials (later compared to numerical results with resistive layer enabled). Stresses are obtained similarly, but from three dimensional equations wherein the temperature profile is used as input. Isotropic materials and small strains are assumed. This means that the model will overestimate temperature/stresses by some margin while maintaining orders of magnitude.

The initial conditions are 293.15 K uniform temperature field and stress-free material. The heat transfer problem is then solved to yield the following temperature function $T(r,t)$ and subsequently stress profiles $\sigma(r,t)$ via thermoelasticity equations

$$T(r,t) = \frac{g}{2a^2}\left((2a^2t + r^2)\cdot erfc\left(\frac{|r|}{2a\sqrt{t}}\right) - 2a|r|\sqrt{\frac{t}{\pi}}\cdot exp\left(\frac{r^2}{4a^2t}\right)\right) \quad (1)$$

$$\sigma_{rr}(r,t) = -\frac{2f}{r^3}\cdot F(r,t) \quad (2)$$

$$\sigma_{\theta\theta}(r,t) = \sigma_{\phi\phi}(r,t) = f\left(\frac{F(r,t)}{r^3} - T(r,t)\right) \quad (3)$$

$$F(r,t) = \int_0^r r^2 T(r,t)dr \quad (4)$$

Here $g, a(r), f(r)$ contain electric, thermal and mechanical parameters of materials and are assumed to be constant in time and space within respective domains. In this case, the above mentioned grain cell dimension is 5.5 μm, where the heat source is at $r = 0$ μm, grain center is at $r = -0.5$ μm, cell boundary is half the grain separation distance away from origin, at $r = 5$ μm.

3.2. Numerical models

COMSOL Multiphysics 5.3 environment is used for finite element modelling throughout this work. Three modules are utilized: AC/DC (solves for electric potential), Heat Transfer (temperature) and Structural Mechanics (displacement field), all of which are coupled. These compute current distribution and Joule power output over the domain of interest, resulting temperature and stresses, as well as other derived physical quantities. Meshes are very fine ($0.005 - 0.02$ μm element size) near the grain boundary and are coarsened further away (element growth rate is tailored for each model). Lines indicated by (x) in figure 2 designate mesh refinement areas that are adapted to channel width changes.

The first model is the toroid heat source along grain equator (Fig. 2.1). The source is artificial: here it is not calculated from current distribution, but rather its intensity matches that of the analytical model. Toroid diameter is set based on r_c/r_g value from the analytical model. Computation results are then compared against analytic estimates to check for consistency.

The next step is the actual discharge model (Fig 2.2). Here current influx is set at the upper end of the channel whereas the other one is grounded. This model comes in two variations with/without material boundary layer. This is used to investigate material boundary layer influence on stress accumulation. Again, to verify model validity, the one with no boundary

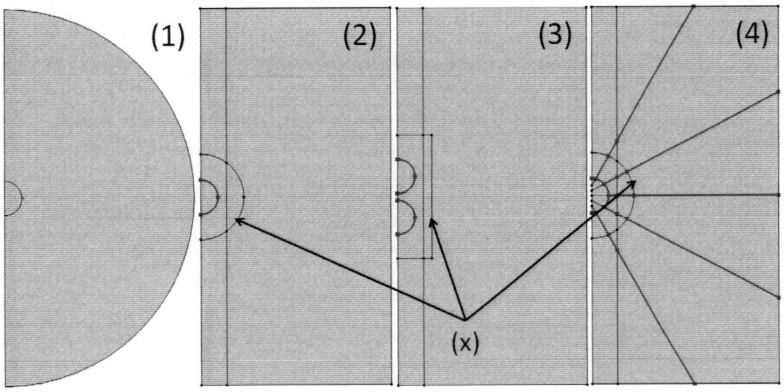

Figure 2. Numerical model geometric configurations: here (k) denotes k-th model geometry (described and referenced in text); hemispheres correspond to carbide grain domains; blue lines represent grain/matrix and martensite sheet boundaries; lines indicated by (x) designate extra mesh refinement areas; vertical lines in (2-4) represent discharge channel boundaries.

layer is parametrically studied to check whether or not and how strongly it is related to the above verification models. Specifically, current channel radius is varied to see if the discharge model converges to the toroid source model as channel radius tends to that of the grain.

Afterwards, this model is further upgraded to three variations: two different configurations where a defect consisting of two adjacent grains is modelled (Fig 2.3 and its counterpart configuration, which is 3 dimensional; see Fig. 1) to investigate proximity effects (grain separation distance is varied); the other is an attempt at martensitic structure modelling (Fig 2.4) where martensite sheets [1] are placed in proximity to the grain and respective boundary layers are accounted for.

Model 1 is 2-dimensional, axially (left boundary) symmetric. The outer boundary is electrically/thermally insulated and set as freely mechanically displaceable. This also applies to models 2, 3, 4. In 1, the heat source is set as a circle of a very small radius. Models 2, 3, 4 include a discharge channel (a narrow rectangle of variable radius with grains inside) with an electrically insulated boundary. Model 3 contains two grains (separation distance variable). Model 4 represents an attempt to model the interaction of grains with martensitic steel sheet shaped domains; specifically, it models the effect of physical contact on stress accumulation near the carbide grain. For these reasons the problem is reduced to an axially symmetric 2-dimensional. The five long radiating lines in Fig 2.4 indicate sheet boundaries.

In addition, inclusion shape study is conducted using model 2. The assumed grain shape is elliptic and its semi axes are varied at constant grain volume for various channel widths to investigate how this affects stress accumulation. There is also the matter of various common defect types in bearing steels aside from the above mentioned carbides. Among these are cavities (voids), $Al_2O_3, MnS, TiCN$ and other types of inclusions [1]. Using model 2, material study is conducted to identify the most dangerous inclusion type. Material parameters for inclusions are taken from Juvonen, P., page 16 [40].

4. Results and analysis

First, the analytical model is used to estimate temperature/stress distributions within the grain and the surrounding steel (figure 3).

Note that the temperature difference (from here on simply temperature) and von Mises stress (from here on simply stress) maxima are just outside the grain boundary. The model clearly indicates that current induced overheating and stress levels are minuscule, about 80 μK and

IOP Conf. Series: Materials Science and Engineering **355** (2018) 012017 doi:10.1088/1757-899X/355/1/012017

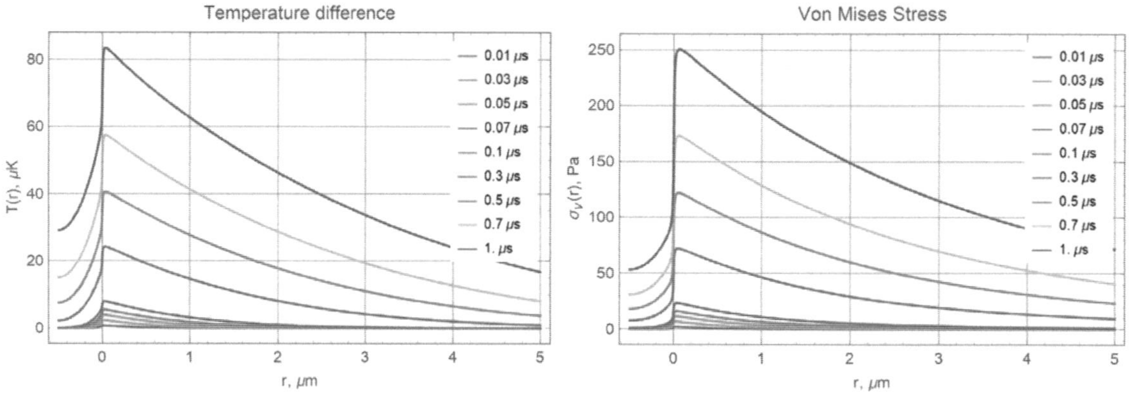

Figure 3. Analytical solution: temperature difference (left) and von Mises stress (right). Grain boundary is at 0 μm.

250 Pa respectively. Stress is predicted to be compressive with respect to the grain. Obviously, profile shapes farther away from the grain surface should be perceived as simple sketches as magnitudes thereof are expected to fall off more rapidly with distance.

Next, the model is used to determine the effective heat impulse propagation time to the adjacent cell boundary. This is important because it indicates whether or not adiabatic external boundary conditions are well suited for numerical models in terms of potentially introduced errors. This characteristic time scale was found to be $\tau = 1.5$ μs, which exceeds the discharge duration. Given the temperature overestimations introduced by the model, this value has a safety margin. Thus, numerical model dimensions will match those proposed in the above sections without risking considerable errors due to heat impulse rebound. This also justifies the use of mechanically free external boundary in numerical models. Nevertheless, all numerical models were also tested by doubling exterior dimensions: the results differed at most by 2%.

As a consistency check numerical model 1 is run, indicating that COMSOL simulations are in order of magnitude agreement with the analytical calculations (figure 4).

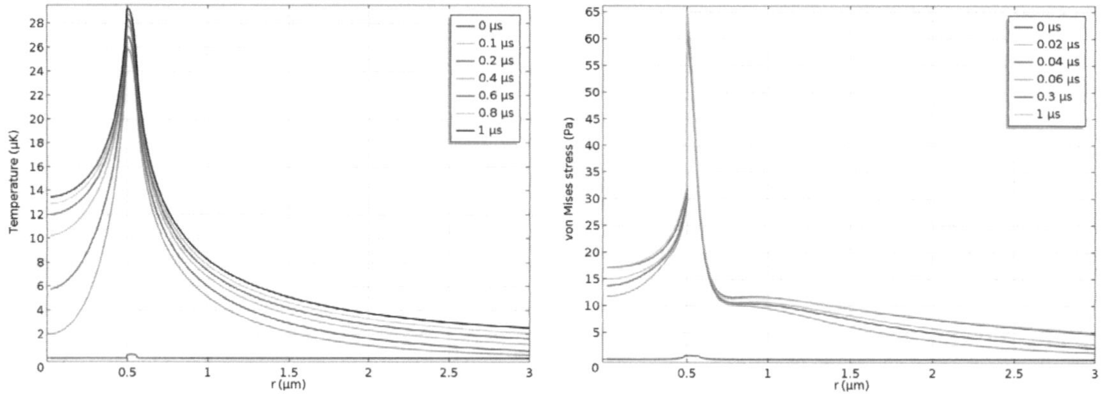

Figure 4. Numerical model 1 results. Grain boundary is at 0.5 μm. Temperature (left) and stress (right) are plotted along heat source coplanar grain radial line.

The rates of temperature/stress build-up are also consistent, given the differences. Temperature profile asymmetry is also quite similar to that predicted analytically, with the expected difference in magnitude decrease over distance. Temperature build-up and stress are concentrated just outside the grain as predicted. Clearly, for such small temperature variations and stresses, the temperature dependence of σ and thermoelastic damping can be neglected and the small strain approximation is valid. Nevertheless, all numerical models are tested and indicate that these effects are in fact irrelevant.

Model 2 is first run for a single grain with/without a boundary layer. Initial channel width is set to $d_c = 1.5\ \mu m$ (the default value for subsequent models). Current is shown to concentrate near grain equator, as initially thought (also seen in [3]), forming a heat source similar in shape to a toroid, which hints at a relationship between this model and the two described above. Indeed, varying d_c shows that as it tends to that of the grain the model converges to the toroid model in the sense that temperature and stress profiles become increasingly similar, almost to the point of matching in shape and magnitude.

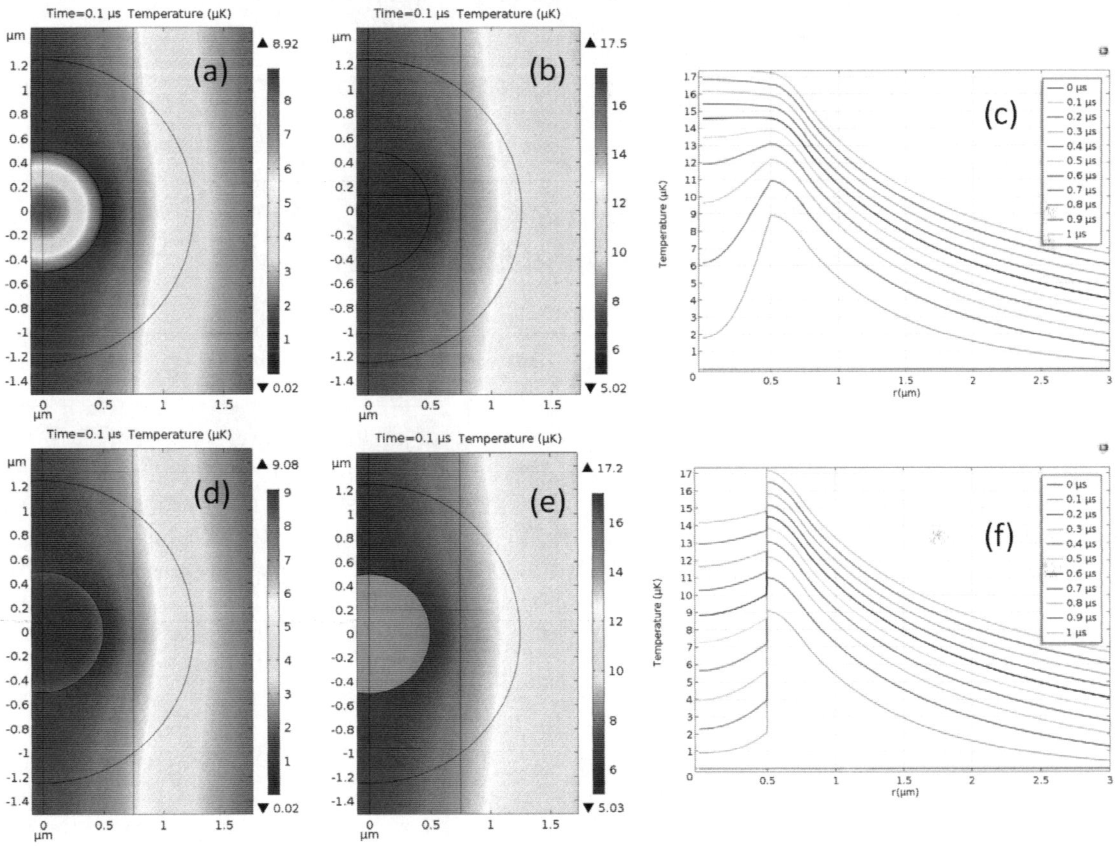

Figure 5. Model 2: temperature field (a,b,d,e) and radial line temperature (c, f). Boundary layer is present in (d,e,f).

It is found that even a very thin material boundary layer has a considerable impact on temperature field evolution for such time scales as evident from figure 5. In absence of the boundary layer the grain quickly overheats, but in the opposite case there is a very steep gradient about the boundary (figure 5c, 5f). It should be noted that figure 5f closely resembles the analytical model solution. This is no coincidence, since solving for two domains separately has the same effect as placing an infinitely resistive thin layer inside the two material interface and more heat is transported further away from the grain, reducing the rate of temperature

IOP Conf. Series: Materials Science and Engineering **355** (2018) 012017 doi:10.1088/1757-899X/355/1/012017

decrease over distance.

Figure 6 Displays the resulting stress field. Material boundary layer presence results in a $\sim$ 10% stress magnitude increase. Note the two regions above and below the grain that are in contrast with the high stress zone. These correspond to stagnation zones where very little current flows.

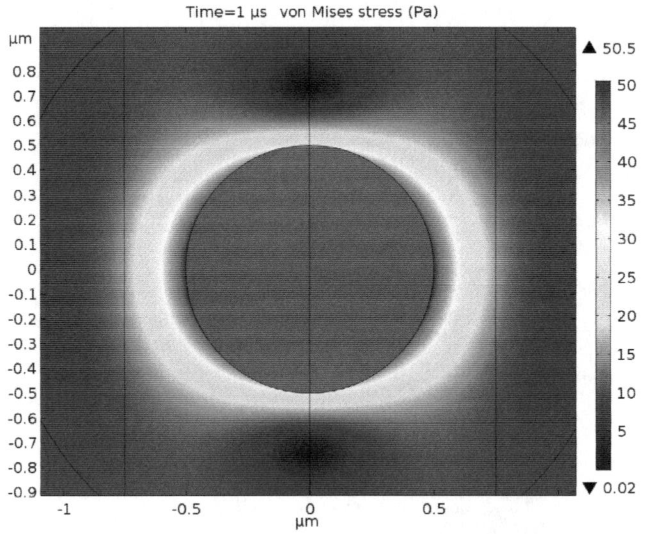

Figure 6. Model 2: stress field in presence of the boundary layer.

Simulations indicate that the grain is generally being stretched along channel axis and compressed in axis orthogonal plane containing the grain equator. As in the above models, stresses monotonously increase over time. Calculated volumetric strain and stored elastic energy values are negligible.

Varying d_c $(1 - 2\ \mu m)$ with fixed total current or fixed current density shows that in the former case narrower channels produce higher stresses, while in the latter the outcome is the opposite, with $\sim$ 20% changes in stress magnitude. All tests result in stress patterns similar to the default (figure 6) case. Boundary layer parameter variations indicate strong sensitivity to its thermal resistance. From here on, all models are implemented with material boundary layers enabled.

Next, model 3 is run. Current flow stagnation zone is observed between the grains, but otherwise the pattern around each grain is similar to the single grain case. Proximity effect manifests as additional stress accumulation (figure 7) due to heat entrapment between the two grains (maximum stress values are still just outside the grain boundary near grain equator).

Grain separation distance parametric study indicates that the maximum stress magnitude increases twofold as the distance decreases from the initial 0.5 μm to 0 μm. Here the grains are stretched in opposite directions away from the proximity zone and compressed in every other direction.

The model where the two grains are placed in the channel cross section plane yield similar results, where extra stress is accumulated between the grains due to current density relative increase as well as proximity effect. Here stress values are $\sim$ 15% greater.

Finally, model 4 is run. It evident that sheet/sheet boundaries act to confine the current flow (figure 8a) to an increasingly narrow channel near the grain boundary. The current is forced through a very small volume close to grain/sheet boundaries (figure 8b). One must note that

IOP Conf. Series: Materials Science and Engineering **355** (2018) 012017 doi:10.1088/1757-899X/355/1/012017

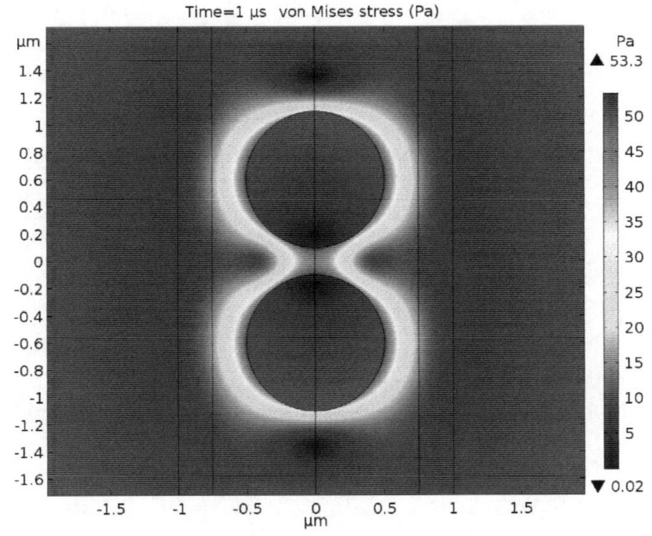

Figure 7. Model 3: stress field resulting from grain proximity. Grain separation is 0.2 μm; $d_c = 1.5$ μm; current flows from top to bottom.

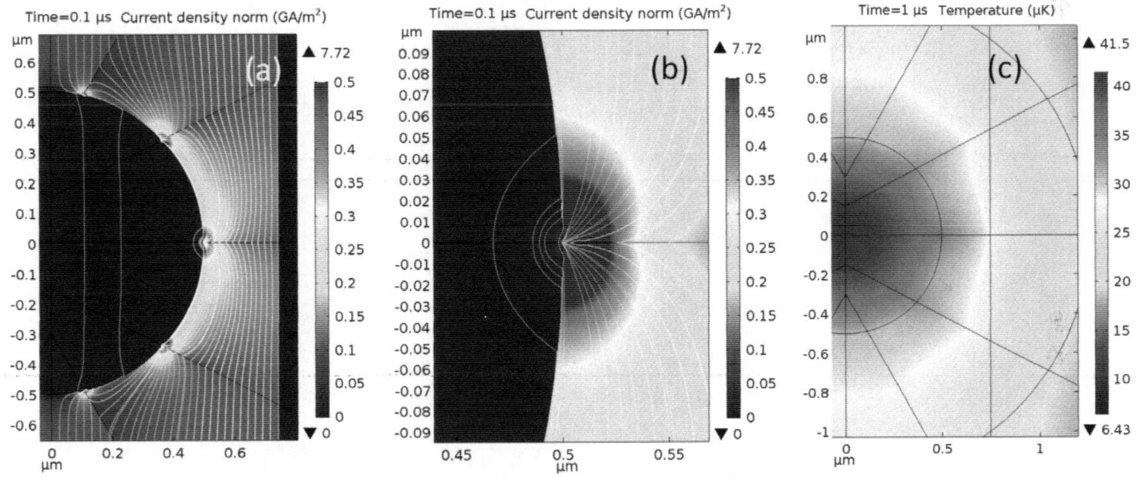

Figure 8. Model 4: current distribution (a, b), temperature field (c). Note: the radial lines within the grain are geometry drawing artefacts, not domain boundaries.

this is only true for material boundary layer σ values less than or close (in figure 8 equal) to that of the carbide grain, as indicated by parametric studies. This results in a number of intense heat generation zones near and beneath the grain surface, especially in the latter case, where, as a consequence, grain volume overheat relative to the exterior steel matrix is very pronounced: the temperature maximum is within the grain volume (figure 8c, 9a), as opposed to the previous cases.

Stress is still concentrated just outside of the grain surface (figure 9b), but magnitudes are up to $\sim$ 40% higher than in the case of homogeneous steel structure. Grain deformation character is the same as in model 2.

Discharge channel width parametric sweep reveals tendencies similar to those observed above,

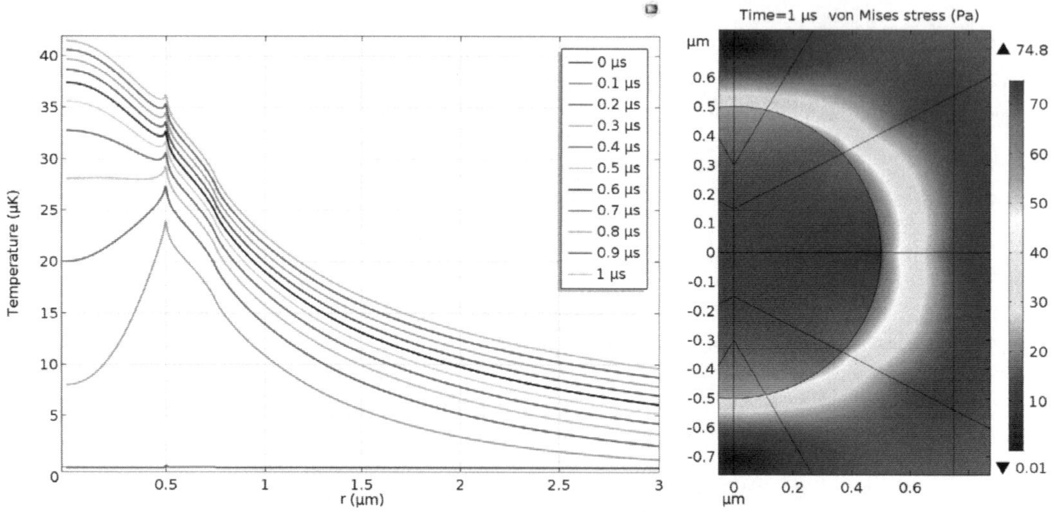

Figure 9. Model 4: radial line temperature (a) and stress field (b).

but unlike the two previous models, this one is much less sensitive to current channel radius variations.

Since $t_{del} = 100\ \mu s$, stress relaxation occurs well before another discharge occurs, as indicated by the frequency-transient analysis of presented models. It is also clear that should the t_{dis}/t_{del} ratio be high enough, residual electrothermal stress accumulation could, though unlikely, eventually assist RCF induced material failure. This is something that needs to be investigated further however, different models must be constructed for larger time scales.

Upon request, Scepanskis, M. has provided the model used in [17] for examination (see figure 11).

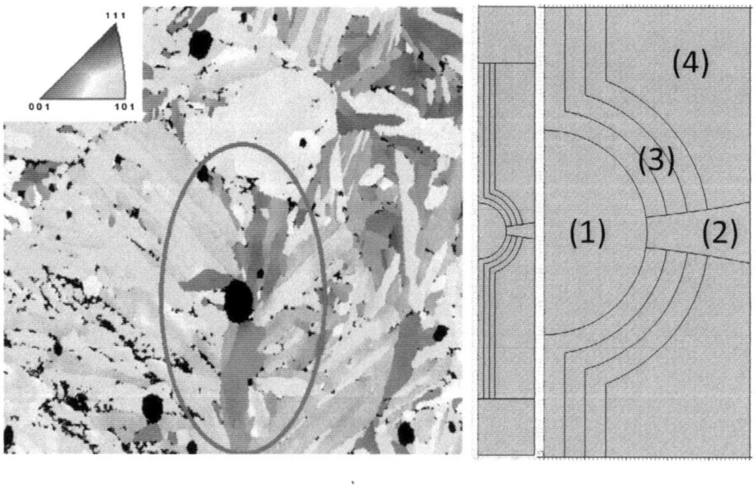

Figure 10. Geometry of the numerical model courtesy of Scepanskis, M. [17]: here (1) is the carbide grain, (4) represents the steel matrix, (2) and (3) designate martensite sheet/needles respectively. The red line (left side) encompasses the microstructure area that the model mimics.

That model is of similar nature, but its radial dimension is 1 μm and $t_{dis} = 100\ \mu s$ by assumption. In figure 11, (1) is the grain ($r_g = 0.5\ \mu m$), (4) is bulk steel matrix: these have the properties given in Tab. 1; (3) represents martensite phase steel needle structure (actually a sheet cross section) [1] and (2) is a domain where orientation differs from (3) and has lower σ and k. It is stated in [17] that C_p of (3), C_{p3}, is expected to be much less than that of (1), (2), (4) and is set to 1 $J/(kg \cdot K)$ (see [17, Fig. 6]). This is not backed by any valid arguments and is dubious given that for temperature range of interest C_p and k of metals are proportional and thus also proportional to σ via the Wiedemann-Franz law (here σ_3 and k_3 are very high, but C_{p3} is very low) and the dimensions of (3) domains are not such that thermal/electrical conduction are restricted enough to affect C_p so drastically. C_p is varied from 1 $J/(kg \cdot K)$ to 100 $J/(kg \cdot K)$, but resulting stresses are not displayed [17].

Thorough testing indicates that certain effects reported in [17] are a product of artefacts introduced by a very coarse mesh. Moreover, the deformations, although claimed to be significant, are not. Volumetric strain calculations yield values of the order of 10^{-3}, which is well within the small strain model. The deformations seen in [17, Fig. 5,6] are found to be a product of scaling by a factor of $\sim$ 5800 (much greater in case of [17, Fig. 7]), which was not mentioned. The size of the reported nanovoid is three orders of magnitude below that of the smallest finite element. Tests with a high quality mesh indicate that for $C_{p3} > 4\ J/(kg \cdot K)$ stresses decrease tenfold and for $C_{p3} > 40\ J/(kg{\cdot}K)$ the model quantitatively reverts to model 2 presented herein; for $C_{p3} > 300\ J/(kg \cdot K)$, the model breaks down, because its $t_{dis} \gg \tau$. Interestingly, this is where the model quantitatively converges to analytical estimates at $C_{p3} = 400\ J/(kg \cdot K)$ both predict stresses on the order of 0.01 to 0.1 MPa, the former yielding greater magnitudes due to heat impulse rebound at the model boundary.

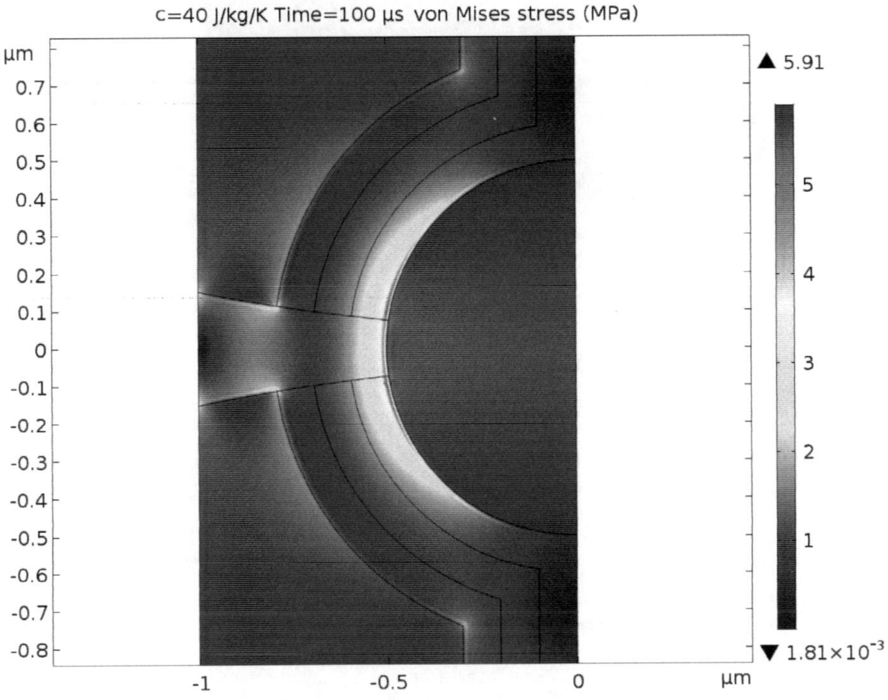

Figure 11. A reproduction of the numerical model courtesy of Scepanskis et al. [17]: stress field for $C_{p3} = 40\ J/(kg \cdot K)$. Note the similarities between this and figures 6, 9b.

As for the inclusion shape/types, it is abundantly clear that varying these does not result in stresses even close to the fatigue limit/yield strength either. It is, however, interesting and important to study the underlying physical picture itself so as to predict inclusion behaviour under similar conditions in the future.

Carbide shape study shows that elongating the grain along the discharge channel axis (case 1) or in the channel cross section plane (case 2) is not equivalent to varying the channel width, as one might expect. For equal grain/channel wall spacing it is found that stresses are always higher in case 2, which is due to the smaller curvature radius of the inclusion boundary near the stress maximum area; there is also considerable inclusion internal stress, as opposed to a spherical grain. However, for a 1 μs discharge, stresses decrease over time in case 2, which can be explained by the fact that the current flow stagnation zones are much larger now and as heat flows away from sources and into these zones, thermal stress relaxation occurs as temperature gradient decreases. In case 1, the dynamics and stress magnitudes are in general similar to what is observed for a spherical grain.

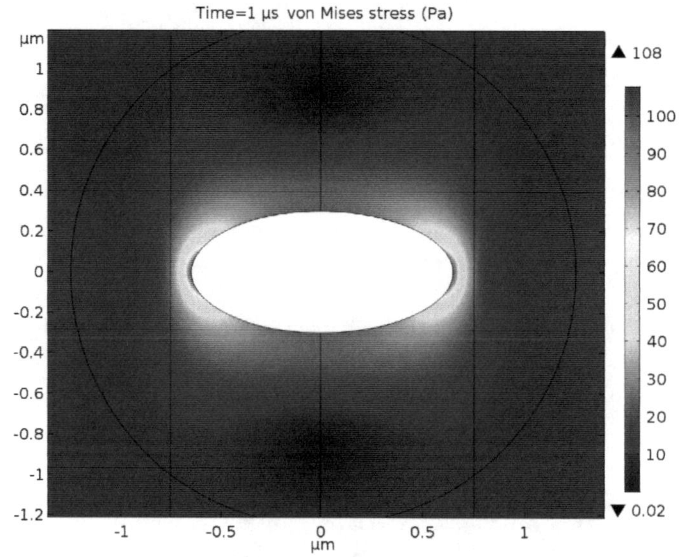

Figure 12. Stress field around an elongated void defect.

Inclusion type study indicated that the most dangerous (in terms of thermal stresses) inclusion type is a void/cavity in steel structure, for which stresses were much higher compared to the carbide grain (figure 12). In fact, for a cavity shaped as in case 2 above, peak stress value is more than double that around the spherical carbide. Ranking the defect types, the carbide grain comes in second, then in, order of descending hazard, come $Al_2O_3, MnO, TiCN$.

5. Conclusions

- It is demonstrated that for discharges of magnitudes/durations modelled herein the resulting stresses are several orders of magnitude below the fatigue limit/yield strength. It is clear then that premature bearing failure observed by Loos et al. and in other similar tests are not due to the mechanisms proposed in present work, especially given that current densities were, most likely, greatly exaggerated.
- Several noteworthy effects are observed: carbide grain/steel material boundary layer noticeably facilitates stress accumulation at inclusions; grains proximity within the discharge channel greatly increases stress levels at said defects; martensitic structure simulations reveal subsurface heat generation within carbide grains in cases where

grain/steel boundary electrical conductivity is close to or lower than that of the grain interior.

- Results obtained by Scepanskis, M. et al. are shown to have been a product of an unfounded assumption about material thermodynamic properties and numerical model definition and implementation issues.
- It is found that the most dangerous kind of inclusion, by a large margin, is a void in steel structure. Then, in order of descent, come carbide, Al_2O_3, MnO and $TiCN$ inclusions. Inclusion shape studies have hinted that elongated defects, when oriented normally to the current flow, tend to be much better stress concentrators that spherical grains, especially in case of structural voids.
- This work is to be perceived as a semi-quantitative study as well, due to many of the parameter values being unknown or known only to a certain degree of approximation (current density, discharge duration/delay, boundary layer parameters, discharge channel geometry).
- New models must be developed in order to investigate transient response of many-defect bearing steel structure for higher t_{dis}/t_{del} values; energy dissipation at material boundaries and electromagnetic forces are to be accounted for all this cannot be accurately done within the framework assumed herein.

6. References

[1] Evans M-H 2016 An updated review: white etching cracks (WECs) and axial cracks in wind turbine gearbox bearings *Mater. Sci. Technol.* **836**
[2] Gegner J 2011 Tribological aspects of rolling bearing failures *Tribology - Lubricants and Lubrication* pp 33-94
[3] Ruellan A 2014 Tribological analysis of white etching crack (WEC) failures in rolling element bearings *PhD thesis*
[4] Evans M-H, Richardson A, Wang L and Wood R 2013 Serial sectioning investigation of butterfly and white etching crack (WEC) formation in wind turbine gearbox bearings *Wear* **302 (s 12)** pp 1573-1582
[5] Evans M-H, Richardson A, Wang L, Wood R and W B Anderson 2014 Confirming subsurface initiation at non-metallic inclusions as one mechanism for white etching crack (WEC) formation *Tribol. Int.* **75** pp 87-97
[6] Gould B and Greco A 2016 An analysis of premature cracking associated with microstructural alterations in an AISI 52100 failed wind turbine bearing using X-ray tomography *Mater. Des.* **117**
[7] Becker P C 1981 Microstructural changes around non-metallic inclusions caused by rolling-contact fatigue of ball-bearing steels *Met. Technol* **8(1)** pp 234-243
[8] Solano-Alvarez W and Bhadeshia H 2014 White-etching matter in bearing steel. Part II: distinguishing cause and effect in bearing steel failure *Metall. Mater. Trans. A* **45(11)**
[9] Brizmer V, Pasaribu H R and Morales-Espejel G 2013 Micropitting performance of oil additives in lubricated rolling contacts *Tribol. Trans.* **176(1-2)**
[10] Kotzalas M N and Doll G 2010 Tribological advancements for reliable wind turbine performance *Philos. Trans. R. Soc. Lond. Ser. A: Math. Phys. Eng. Sci.* **368(1929)** pp 4829-4850
[11] Fitch J 2008 How water causes bearing failure *Machinery Lubrication Magazine*
[12] Szost B and Rivera-Diaz-del-Castillo P 2013 Unveiling the nature of hydrogen embrittlement in bearing steels employing a new technique *Scr. Mater.* **68** pp 467-470
[13] Nagumo N 2004 Hydrogen related failure of steels - a new aspect *Mater. Sci. Technol.* **20(8)** pp 940-950
[14] Birnbaum H K and Sofronis P 1994 Hydrogen-enhanced localized plasticity - a mechanism for hydrogen-related fracture *Mater. Sci. Eng., A* **176(1-2)** pp 191-202
[15] Oriani R A 1987 Hydrogen - the versatile embrittler *Corrosion* **43(7)** pp 390-397
[16] Kang Y S, Evans R and Doll G 2010 Roller-raceway slip simulations of wind turbine gearbox bearings using dynamic bearing model *STLE/ASME 2010 International Joint Tribology Conference*
[17] Scepanskis M, Jakovics A, Kaldre I, Holweger W, Nacke B and Diederichs A-M 2015 The numerical model of electrothermal deformations of carbides in bearing steel as the possible cause of white etching cracks initiation *Tribol. Lett.* **59(2)** pp 1-10
[18] Loos J, Bergmann I and Goss M 2016 Influence of currents from electrostatic charges on WEC formation in rolling bearings *Tribol. Trans.* **59(5)**

[19] Prashad H 1999 Determination of magnetic flux density on the surfaces of rolling-element bearings as an indication of the current that has passed through them - an investigation *Tribol. Inter.* **32(8)** pp 455-467.
[20] Bruce T, Long H and Dwyer-Joyce R S 2015 Dynamic modelling of wind turbine gearbox bearing loading during transient events *IET Renew. Power Gen.* **9(7)** pp 821-830
[21] Rosinski J and Smurthwaite D 2010 Troubleshooting wind gearbox problems *Wear* pp 22-33
[22] Smelova V, Schwedt A, Wang L, Mayer J and Holweger W 2017 A study on evolution of microstructure alteration in white etching crack *Int. J. Fatigue* **100**
[23] Gould B and Greco A 2016 Investigating the process of white etching crack initiation in bearing steel *Tribol. Lett.* **62(2)**
[24] Evans M-H 2013 White structure flaking failure in bearings under rolling contact fatigue *PhD thesis*
[25] Holweger W 2014 Progress in solving white etching crack phenomena *Gearbox reliability collaborative all member meeting, National Renewable Energy Laboratory*
[26] Diederichs A-M, Schwedt A, Mayer J and Dreifert T 2015 Electron microscopy analysis of structural changes within white etching areas *J. Mater. Sci. Technol.*
[27] Iso K, Yokouchi A and Takemura H 2005 Research work for clarifying the mechanism of white structure flaking and extending the life of bearings *SAE 2005 World Congress and Exhibition*
[28] Scepanskis M, Gould B and Greco A 2017 Empirical investigation of electricity self-generation in a lubricated sliding-rolling contact *Tribol. Lett.* **65(3)**
[29] Landau P, Osovski S, Venkert A, Gartnerova V and Rittel D 2016 The genesis of adiabatic shear bands *Scientific Reports 6 - Nature*
[30] Wright T W 2002 The physics and mathematics of adiabatic shear bands *Cambridge University Press*
[31] Kadin Y and Sherif M Y 2016 Energy dissipation at rubbing crack faces in rolling contact fatigue as the mechanism of white etching area formation *Int. J. Fatigue* **96**
[32] Konrad A, Nierlich W and Gegner J 2013 Finite element analysis of the rolling-sliding contact of vibrationally loaded bearings based on a micro friction model *Mater. Sci. Forum* **768-769** pp 714-722
[33] Moghaddam S M, Sadeghi F, Weinzapfel N and Liebel A 2014 A damage mechanics approach to simulate butterfly wing formation around nonmetallic inclusions *J. Tribol.* **137(1)**
[34] Kang J-H and Rivera-Diaz-del-Castillo 2013 Carbide dissolution in bearing steels *Comput. Mater. Sci.* **67** pp 364-372
[35] Nishimura K and Miyazaki N 2004 Molecular dynamics simulation of crack growth under cyclic loading *Comput. Mater. Sci.* **31(3-4)** pp 269-278
[36] Ackland G J, Jones A P and Noble-Eddy R 2008 Molecular dynamics simulations of the martensitic phase transition process *Mater. Sci. Eng., A* **481(1)** pp 11-17
[37] Grujicic M, Chenna V, Galgalikar R, Snipes J S, Ramaswami S and Yavari R 2014 Wind-turbine gear-box roller-bearing premature-failure caused by grain-boundary hydrogen embrittlement: a multi-physics computational investigation *J. Mater. Eng. Perform.* **23(11)** pp 3984-4001
[38] Grujicic M, Ramaswami S, Yavari R, Galgalikar R, Chenna V and Snipes J S 2014 Multiphysics computational analysis of white-etch cracking failure mode in wind turbine gearbox bearings *Proc. Inst. Mech. Eng. L: J. Mater. Des. Appl.* **230(1)**
[39] Winzer N and Khader I 2013 Hydrogen diffusion and trapping in bodies undergoing rolling contact *Wear* **303(s 12)** pp 451-458
[40] Juvonen P 2004 Effects of non-metallic inclusions on fatigue properties of calcium treated steels *PhD dissertation* p 16

IOP Conf. Series: Materials Science and Engineering **355** (2018) 012018 doi:10.1088/1757-899X/355/1/012018

Thermo-Chemical Conversion of Microwave Activated Biomass Mixtures

I Barmina[1], A Kolmickovs[1], R Valdmanis[1], S. Vostrikovs[1] and M Zake[1]

[1] Institute of Physics, University of Latvia, Salaspils, 32 Miera St, LV-2169, Latvia

E-mail: inesa.barmina@lu.lv

Abstract. Thermo-chemical conversion of microwave activated wheat straw mixtures with wood or peat pellets is studied experimentally with the aim to provide more effective application of wheat straw for heat energy production. Microwave pre-processing of straw pellets is used to provide a partial decomposition of the main constituents of straw and to activate the thermo-chemical conversion of wheat straw mixtures with wood or peat pellets. The experimental study includes complex measurements of the elemental composition of biomass pellets (wheat straw, wood, peat), DTG analysis of their thermal degradation, FTIR analysis of the composition of combustible volatiles entering the combustor, the flame temperature, the heat output of the device and composition of the products by comparing these characteristics for mixtures with unprocessed and mw pre-treated straw pellets. The results of experimental study confirm that mw pre-processing of straw activates the thermal decomposition of mixtures providing enhanced formation of combustible volatiles. This leads to improvement of the combustion conditions in the flame reaction zone, completing thus the combustion of volatiles, increasing the flame temperature, the heat output from the device, the produced heat energy per mass of burned mixture and decreasing at the same time the mass fraction of unburned volatiles in the products.

1. Introduction

The EU 2020 strategy targets on climate change and energy, which comply with the Kyoto Protocol, prescribe the increased use of cleaner renewable energy sources (by 20%), partially replacing the fossil fuels with the second-generation fuel – wood and agriculture residues [1] with growing demand for application of agriculture residues for energy production [2–5]. The main advantage of using straw (wheat straw, rape straw) as fuel in the energy sector - straw is carbon (CO_2) neutral fuel and does not contribute to an increase of greenhouse gas (GHG) emissions from energy producers. However, straw has been known as a more problematic fuel if compared with wood because of its lower heating values (LHV, HHV), higher nitrogen and ash contents in biomass [6, 7]. To ensure wider use of straw for energy production, co-combustion of straw with renewable or fossil fuels is used, reducing the greenhouse gas emissions during the heat energy production and improving the main combustion characteristics compared to straw [8-12]. Moreover, it has been demonstrated that an interesting and efficient alternative for biomass conversion to high quality biofuel is mw-pretreatment of biomass with direct influence on the main gasification/combustion characteristics and composition of the products [13–17]. The results of previous research confirm that mw pre-processing of biomass pellets reduces their moisture content and partially decomposes hemicelluloses, cellulose and lignin, decreasing the hydrogen-to-carbon (H/C) and oxygen-to-carbon (O/C) contents while increasing calorific value of pellets. This leads to faster and more complete thermo-chemical conversion of pre-processed pellets by

Content from this work may be used under the terms of the Creative Commons Attribution 3.0 licence. Any further distribution of this work must maintain attribution to the author(s) and the title of the work, journal citation and DOI.
Published under licence by IOP Publishing Ltd

increasing the produced heat energy per mass of burned biomass pellets [18, 19]. Besides, this prognosis that an additional improvement of the main combustion characteristics at co-combustion of straw with wood or peat pellets can be obtained using the selective microwave (mw) pre-processing of straw pellets with activation of the thermo-chemical conversion of straw mixtures. Hence, the main aim of the current research is to provide the detailed experimental study of the main combustion characteristics at co-combustion of straw with wood or peat pellets providing the selective mw pre-processing of straw pellets to assess and analyze the effect of mw pre-processing on the development of the main gasification/combustion characteristics, heat output of the device and composition of the products at thermo-chemical conversion of activated mixtures.

2. Experimental

The effects of mw pre-processing of the mixture components (wheat straw mixtures with wood or with peat pellets) on the gasification/combustion characteristics were studied using a pilot device (60 mm in diameter and 600 mm in length) with a heat output up to 2 kW. The device combines a biomass gasifier (1) filled with biomass pellets and the water-cooled sections of the combustor (2) (figure. 1).

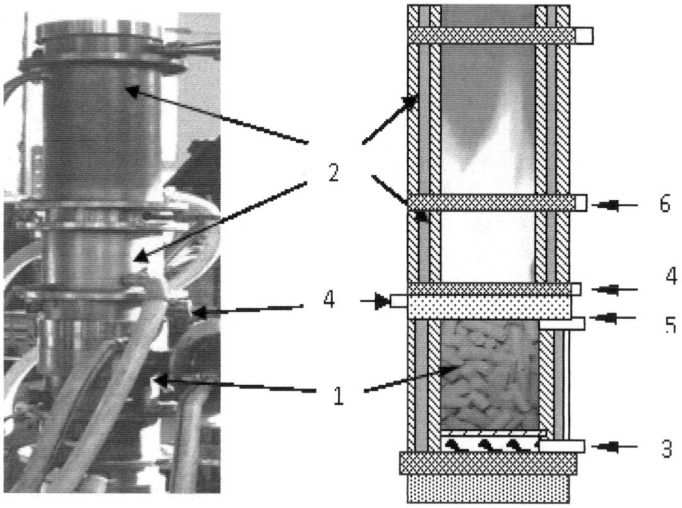

Figure 1. Real-life photo and schematic image of the pilot device:

1 – gasifier of biomass mixture;
2 – sections of the combustor;
3 – primary air supply nozzle;
4 – secondary air supply nozzle;
5 – propane flame supply;
6 – orifice for diagnostic tools.

The experiments were carried out using mixtures of straw with wood or peat pellets. The mass load of straw in the mixture was 30%. To provide activation of the straw thermal decomposition, the straw pellets were pre-processed in a microwave oven with power 700 W and the frequency $\nu = 2.45$ GHz with duration of microwave pre-processing 180 s. The gasification/combustion characteristics were studied experimentally using the primary air (q_1) supply below the biomass layer (3) at the average rate $q_1 = 20$ l/min to sustain the formation of axial flow of combustible volatiles (CO, H_2), which produced during the thermal decomposition of the biomass mixtures. The secondary swirling air flow was supplied at the bottom of the combustor (4) at the average rate $q_2 = 30$ l/min to sustain the burnout of the volatiles downstream the combustor. A propane flame flow was supplied into the upper part of the biomass layer (5) to initiate the thermal decomposition of the biomass pellets. The supply was stopped after ignition of volatiles (t = 360 s).

To assess the influence of straw mw pre-processing on the development of the main gasification/combustion characteristics at thermo-chemical conversion of activated mixtures, the experimental study involved joint measurements of the main characteristics (elemental composition, heating values) of parent samples (straw, wood and peat pellets) and pre-treated straw pellets, DTG and DTA analysis of their thermal decomposition [18] and complex time-dependent measurements of weight loss rates of the mixtures (dm/dt, g/s), the composition of released volatiles entering the combustor, the flame temperature, the heat output from the device and the composition of the products. FTIR spectrum analysis in the mid-IR range is used for measurements of volatiles composition entering the combustor.

The gas sampling probe was inserted through the gas sampling orifice (6). Pt-Pt/Rh thermocouples were used to measure the flame temperature with data online registration by Pico logger. To assess the influence of the straw pre-processing on the heat output of the device, calorimetric measurements of the cooling water flow were made using thermos-sensors AD590 with online data registration using Data Translation DT 98005 series data acquisition module. The composition of the products (CO, H_2, CO_2, NO_x) and the combustion efficiency were measured using a Testo 350 gas analyzer. The biomass weight loss rate was estimated from time-dependent variations of the biomass layer height in the gasifier using a moving rod with a pointer. Online data registration of the main gasification/combustion characteristics was made once per second.

3. Results and discussion

The comparative analysis of the main characteristics of the parent and mw-pre-processed samples of wheat straw, wood and peat pellets has shown a considerable difference of their elemental composition and heating values (table 1). Wood and peat pellets have a higher carbon and hydrogen contents and higher heating values, whereas straw pellets have higher nitrogen and ash content. Therefore, measurements of the main characteristics of biomass pellets suggests that mixing of wheat straw with wood or peat pellets is preferable to create a fuel mixture with improved elemental composition and heating values (HHV) compared to straw (table 1). Therefore, it is believed that co-combustion of straw with wood or peat pellets will allow to improve the main combustion characteristics by increasing the heat output from the device and to improve the products composition so determining wider use of straw for energy production.

Table 1. The elemental composition and heating values (HHV) of unprocessed straw, mw pre-processed straw, wood and peat pellets, and mixtures of both unprocessed (30%) and mw pre-processed straw with wood or peat pellets.

Parameter	wheat straw	wheat straw (mw)	wood	peat	wood + 30% of		peat + 30% of	
					straw	straw (mw)	straw	straw (mw)
C, %	46.62	49.70	49.79	53.83	48.84	49.76	51.66	52.59
H, %	5.09	5.25	5.15	5.12	5.132	5.18	5.11	5.16
O, %	42.72	38.15	44.24	36.93	43.78	42.41	38.66	37.30
N, %	1.31	1.46	0.18	1.11	0.52	0.56	1.17	1.215
Ash, %	4.26	5.44	0.64	3.02	1.73	2.08	3.39	3.75
H_2O^a, %	9.09	1.07	6.32	11.44	7.151	4.745	10.74	8.33
HHV[b]	18.47	19.68	19.52	21.24	19.20	19.57	20.41	20.77

[a] Moisture content.
[b] High Heating Value, MJ/kg.

The results of previous research have shown that additional improvement of the elemental composition and main combustion characteristics of lignocellulosic biomass pellets can obtained at their microwave pre-processing [18, 19], which starts with the removal of the physically bounded water and is followed by the changes in their chemical composition with the breakdown of the glycosidic structure of polysaccharides into smaller fragments, reducing the crystallinity of cellulose and providing carbonization of the cell wall structure [13–16]. Similar structural transformation along with partial destruction of hemicelluloses and cellulose and variations of the elemental composition are observed at microwave pre-processing of wheat straw pellets indicating a decrease of the moisture, hydrogen and oxygen contents in pellets, whereas increases the carbon content and heating values (HHV) of pellets (table 1). The main negative effect of mw pre-processing of wheat straw pellets is an increase of nitrogen and ash contents in pellets by limiting their use as fuel for energy production. To minimize these negative effects, co-combustion of activated wheat straw with wood or peat pellets is still preferable (table 1) suggesting that co-combustion of these pellets will allow to control and improve the elemental composition of fuel mixtures with direct influence on their thermal decomposition and combustion of volatiles.

To predict the thermal decomposition of biomass mixtures, detailed DTG analysis of the main components – wheat straw (parent and activated samples), wood and peat pellets in oxidative atmosphere was carried out indicating that difference in their elemental and chemical composition (table 2) influences their thermal decomposition and the formation of combustible volatiles.

Table 2. Chemical composition of biomass pellets [5, 18].

Biomass sample	Hemicelluloses, %	Cellulose, %	Lignin, %
wheat straw	21 – 28	24 – 35	16 – 20
wood	23 – 25	41 – 43	28 – 29
peat	10 – 25	0 – 20	6 – 40

In accordance with data reported in [20–23] the formation of the weight loss peaks at thermal decomposition of lignocellulosic biomass in oxidative environment can be related to the first step of the low-temperature formation of volatiles and char residue with the next step of lignin thermal decomposition and char conversion. DTG analysis of unprocessed wheat straw thermal decomposition in oxidative atmosphere indicates the formation of the weight loss peaks at T ≈ 560 K, 650 K and at around 711 K (figure 2–a). The detailed analysis of wheat straw thermal degradation suggests [21] that below T < 630 K dominates the thermal decomposition of hemicelluloses and cellulose, whereas at T ≈ 630–670 K dominates the thermal decomposition of lignin, which is highly responsible for char formation and conversion. The results of DTA analysis (figure 2–c) indicate that up to 540 K develop the endothermic processes of the thermal decomposition of hemicelluloses and cellulose, while the exothermic processes of volatile ignition and combustion start to develop at T > 540 K promoting a fast increase of the wheat straw weight loss rate and temperature of pellets up to their peak values, which are detected at T ≈ 560 K. The formation of exothermic peaks at T ≈ 640–650 K and T ≈ 700–710 can be related to combustion of volatiles and char conversion, which is produced at thermal decomposition of wheat straw lignin [21]. The enhanced thermal decomposition of hemicelluloses, cellulose and lignin is observed at microwave pre-processing of wheat straw pellets increasing the weight loss peaks at around 550–560 K and 640–650 K, whereas slightly reduces the weight loss peak at T ≈ 710 K, which can be related to char conversion (figure 2–b).

The pronounced weight loss peaks at thermal decomposition and char conversion in oxidative environment are observed also for wood and peat pellets (figure 2–a). As follows from figure 2–a, the thermal decomposition of wood pellets occurs at higher temperature T ≈ 607 K, because wood pellets have higher content of cellulose and lignin compared to straw (table–2), which decomposes at higher temperature than hemicelluloses producing higher content of volatiles and char than straw. The weight loss peak at T ≈ 720 K for wood pellets can be related to char conversion, which is produced during the thermal decomposition of cellulose and lignin. DTA analysis of wood pellets (figure 2–c) suggests that the formation of weight loss peaks at T ≈ 607 K and T ≈ 716–720 K for wood pellets can be related to combustion of volatiles and char conversion.

The formation of the weight loss peaks at T ≈ 560 K and T ≈ 650 K is observed at the thermal decomposition of peat pellets with more pronounced weight loss peak at T ≈ 650 K, which can be related to char formation and conversion (figure 2–a).

This suggests higher lignin content in peat pellets, which produces more char during their thermal decomposition. The results of DTA analysis in oxidative environment confirm the enhanced char formation and conversion for peat pellets with enhanced increase of the temperature during the char conversion stage. Hence, due to differences of elemental and chemical composition between straw, wood and peat pellets, the thermal decomposition of their mixtures is influenced by the main characteristics of the mixture components, and can be quite different varying the mixture composition, which is confirmed by the kinetic study of the thermal decomposition of straw pellets and their mixtures with wood or peat.

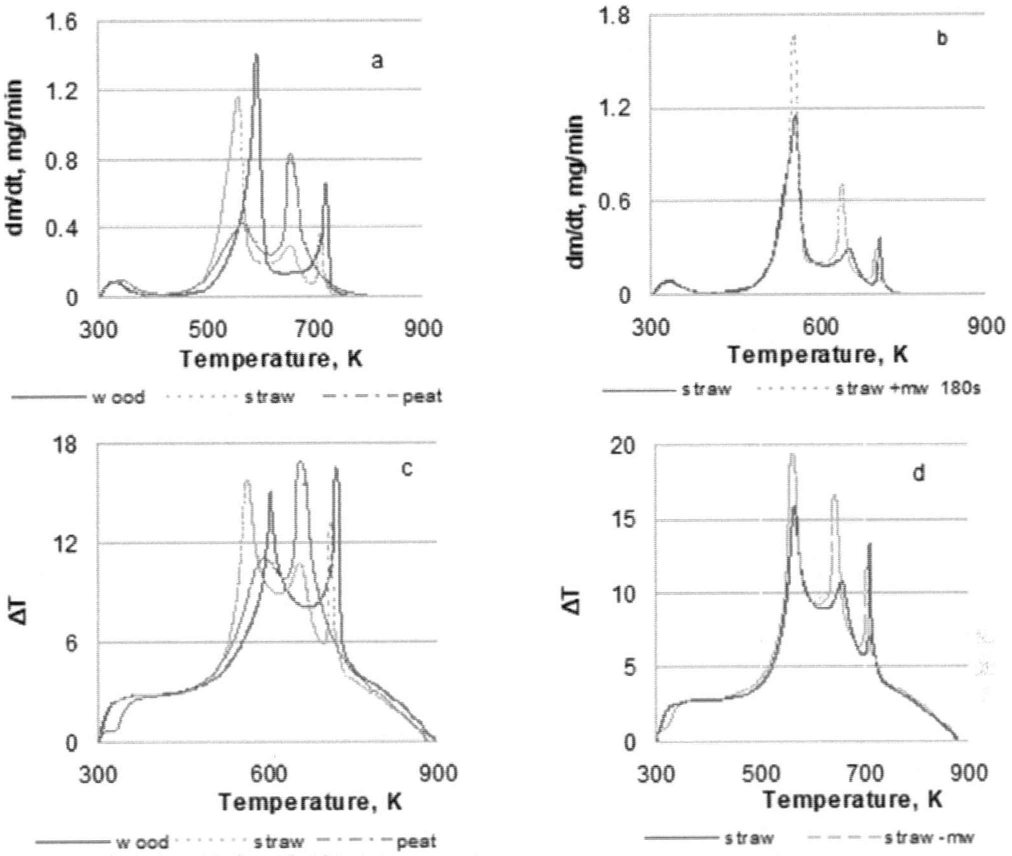

Figure 2. The weight loss rate (DTG) and temperature (DTA) of biomass pellets (a, c) and their variations at mw pre-processing of straw (b, d).

Comparative kinetic study of the of the weight loss rates for unprocessed and mw pre-processed wheat straw pellets at their thermal decomposition developing in the gasifier of the pilot device (figure 1) has shown that microwave pre-processing of wheat straw provides enhanced thermal decomposition of straw pellets. The average weight loss rate increases from 0.1 g/s for unprocessed wheat straw pellets up to 0.12 g/s (by about 20%) at microwave pre-processing of straw. In accordance with data of DTG analysis (figure 2) the formation of a primary sharp weight loss peak for mw activated wheat straw pellets refers to flaming combustion of volatiles ($t = 400–900$ s). The next weight loss peak, the formation of which is observed during the time interval $t \approx 1000–1400$ s, can be related to the enhanced thermal degradation of straw lignin [21] followed by the enhanced formation and combustion of volatiles with char formation and combustion at $t > 1400$ s (figure 3–a).

The results of kinetic study show that mw pre-processing of straw pellets makes an influence on the weight loss rates of activated straw mixtures with wood or peat pellets (figure 3–b, c) increasing their average weight loss rates. The average weight loss of straw mixture with wood pellets increases from 0.11 g/s for unprocessed straw pellets up to 0.13 g/s (≈ 15-16%) at selective mw pre-processing of straw. By co-firing of straw with peat pellets the average weight loss rate of their mixture increases from 0.094 g/s for unprocessed straw pellets up to 0.116 g/s (≈ 23.4%) regarding the selective mw pre-processing of straw. It should be noted that for mixture of activated straw with peat pellets the weight loss of the mixture increases both during the flaming combustion of volatiles and after flame smoldering or char conversion stages. The enhanced thermal decomposition of mixtures at mw-pre-processing of straw pellets correlates with the enhanced thermal decomposition of wood and peat pellets providing faster burnout of activated mixtures (figure 3–b, c). This suggests the thermal interaction between the components, when the enhanced volatilization of pre-processed straw pellets and combustion of volatiles promotes the enhanced heat release with activation of the thermal decomposition of wood or peat pellets and thermo-chemical conversion of their mixtures.

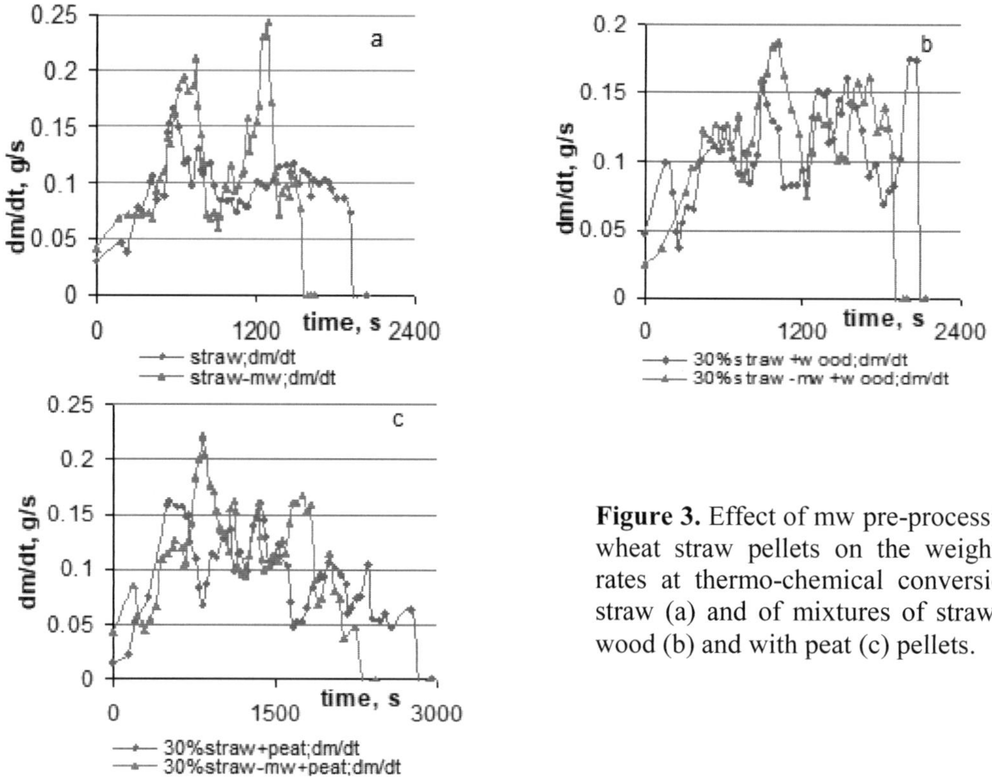

Figure 3. Effect of mw pre-processing of wheat straw pellets on the weight loss rates at thermo-chemical conversion of straw (a) and of mixtures of straw with wood (b) and with peat (c) pellets.

Because of the enhanced thermal decomposition of activated mixtures larger amounts of combustible volatiles are produced. This is confirmed by the results of FTIR analysis of the volatiles composition entering the combustor (figure 4). The FTIR analysis of volatiles composition confirms the enhanced release of combustible volatiles (H_2, CO, C_xH_y) (figure 4–a, b) at thermal decomposition of activated mixtures (figure 4).

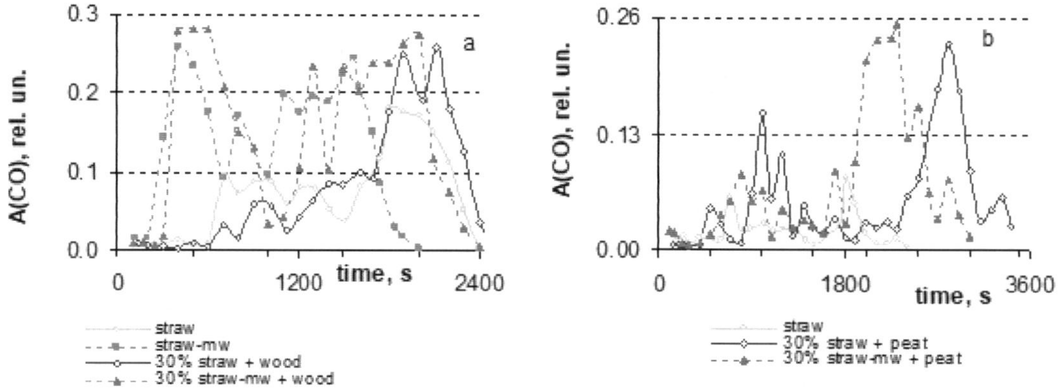

Figure 4. Effect of mw pre-processing of wheat straw pellets on the release of combustible volatiles (CO) at thermochemical conversion of activated straw mixtures with wood (a) and with peat (b) pellets.

When comparing the kinetics of the formation of volatiles for the activated mixtures of straw with wood or with peat pellets, one can see a difference in time-dependent variations of the formation of combustible volatiles. Co-combustion of wood pellets with activated straw resulted in an enhanced release of CO during the primary stage of thermal decomposition of activated mixture (t < 800–1000 s), when in accordance with data of DTG analysis dominates the enhanced thermal decomposition of hemicelluloses and cellulose. In contrary, the mixture of activated straw with peat pellets demonstrated the enhanced release of CO during the after-flame char conversion stage indicating that enhanced thermal decomposition of peat lignin promotes the enhanced char formation and conversion.

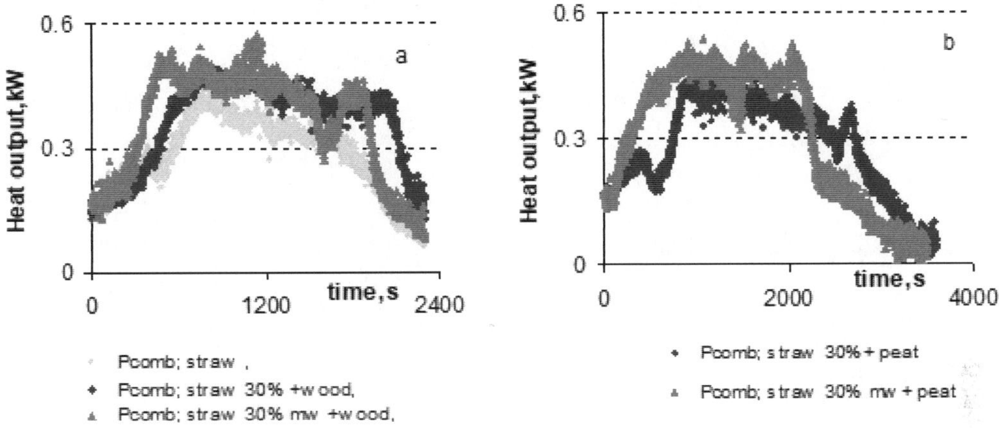

Figure 5. Effect of wheat straw mw pre-processing on the heat output from the flame reaction zone at thermochemical conversion of activated straw mixtures with wood (a) and peat (b) pellets.

Because of the enhanced thermal decomposition of activated mixtures, the mass flow of combustible volatiles entering the combustor increases. For the constant primary and secondary air supply rates into the device, the thermo-chemical conversion of the unprocessed straw mixtures with wood or with peat pellets is developing at the average air excess ratio in the flame reaction zone about $\alpha \approx 1.6$–1.7. Hence, increasing the mass flow of combustible volatiles into the combustor provides a decrease of the air to fuel supply ratio decreasing the average value of the air excess ratio in the flame reaction zone. By co-firing of straw with wood pellets the average air excess ratio at microwave pre-processing of straw decreases from $\alpha \approx 1.7$ to $\alpha \approx 1.4$ demonstrating improvement of the combustion conditions. Similar improvement of the combustion conditions is observed by co-firing of straw with peat pellets, when the average air excess ratio at microwave pre-processing of straw decreases from $\alpha \approx 1.65$ to $\alpha \approx 1.35$ providing faster and more complete combustion of volatiles with correlating increase of the peak and average values of the heat output from the device (figure 5–a, b) and produced heat energy per weight of the burned mixture. The average heat output from the reaction zone at thermo-chemical conversion of the activated straw-wood mixture can be increased by ~25% increasing the total heat output from the device by ~12%. The heat output from the reaction zone for the activated straw-peat mixture grew up by ~13% with the correlating increase of the total heat output from the device by ~20%. The produced heat energy per mass of activated mixtures of straw with wood or peat increases by ~3–6%, which confirms that the mw activation of straw pellets results in more complete thermo-chemical conversion of the mixture. It should be noted however that produced heat energies per mass of activated straw and burned pellets of wood or peat are not additive. This is confirmed by the results of experimental study indicating that produced heat energy per mass of activated mixtures slightly (by about 5–7.3%) exceeds the total amount of heat energy produced at thermo-chemical conversion of activated straw and biomass pellets (wood or peat). This suggests that thermal decomposition of activated mixtures influences the composition of volatiles entering the combustion (figure 4) and reactions developing in the flame reaction zone.

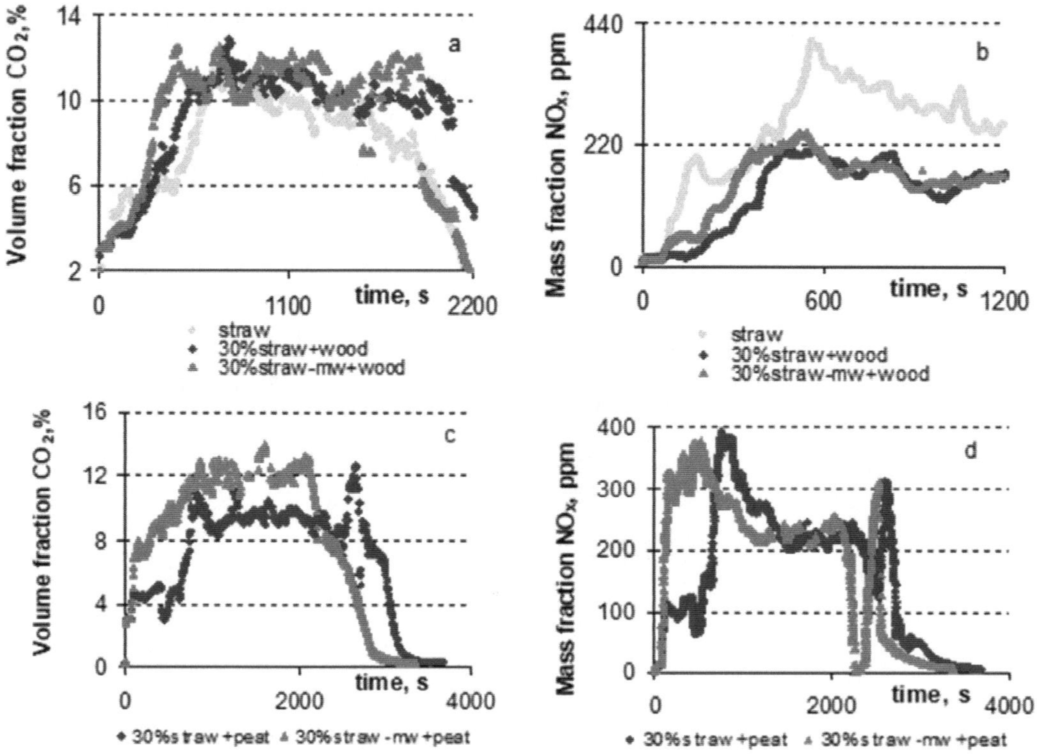

Figure 6. Effect of wheat straw pellets mw pre-processing on greenhouse carbon (CO_2) (a, c) and polluting NO_x emission (b, d) at thermo-chemical conversion of activated straw mixtures with wood (a, b) and with peat (c, d) pellets.

Finally, the activation of the thermal decomposition of mixtures shows the influence on the products composition (figure 6–a, c). At thermo-chemical conversion of activated straw mixtures with wood pellets the average value of the CO_2 volume fraction in the products increases by ~2%. For the mixture of activated straw with peat pellets the average volume fraction of CO_2 in the products increases by ~2.2% with correlating decrease of the mass fraction of volatiles (CO, H_2) in the products. The average mass fraction of CO decreases from 1190 ppm to 350 ppm, whereas the average mass fraction of H_2 in the products decreases from 340 ppm to 270 ppm indicating more complete combustion of volatiles. In addition, the co-combustion of straw with wood at average mass load of straw 30% results in a decrease of the average values of the polluting NO_x emissions from 214 ppm to 190 ppm (by ~10%). A slight growth of the NO_x emission in the products (by ~5%) was observed for the mixture of straw with peat, which predominately can be related to the higher nitrogen content in the peat pellets (1.14%) in comparison with straw (1.01%).

4. Conclusions

The co-combustion of straw with wood or with peat pellets, with the 30% straw mass load in the mixture, was studied experimentally to provide wider use of straw as a fuel with more complete thermo-chemical conversion of pellets, increased heat output from the device and reduced polluting NO_x emission in the products in comparison to the thermo-chemical conversion of straw.

Additional improvement of the combustion conditions and composition of emissions is achieved by mw pre-processing of straw pellets. The activation of straw pellets by mw pre-processing provides enhanced thermal decomposition of the activated straw mixtures with wood and with peat pellets promoting enhanced release and more complete thermo-chemical conversion of the combustible volatiles (CO, H_2) and thus increasing the total heat output from the device by about 12% at the co-combustion of straw with wood and by about 20% at the co-combustion of activated straw with peat.

The enhanced release and more complete thermo-chemical conversion of volatiles result in increase of the average value of the carbon-neutral CO_2 fraction in the products by about 2–2.2%. The thermo-chemical conversion of the activated straw mixture with wood pellets leads to reduction of the mass fraction of polluting NO_x emissions in the products by about 10%, and, in contrary, the emission of NO_x at the thermo-chemical conversion of the activated straw mixture with peat slightly increases (by about 5%) of, which is the main negative result at co-combustion of straw with peat.

References

[1] Oettinger G H 2011 A strategy for competitive, sustainable and secure energy *Energy 2020* (European Union) p 28, https://ec.europa.eu/energy/sites/ener/files/documents/2011_energy2020_en_0.pdf
[2] Skott T 2011 *Straw to energy- status, technologies and innovation in Denmark (*Agro Business Park A/S) p 22 agropark.dk
[3] Glithero N J, Wilson P and Ramsden S J 2013 *Biomass and Bioenergy* **55** 311-321
[4] Larsson S H, Therel M, Geladi P and Lestander T A 2008 *Bioresour.Technol.* **99** 7176-7182
[5] Olsson M 2006 *Biomass and Bioenergy* **30** 555-564
[6] S. Vasilev, D. Baxter, L.K. Andersen, C.G. Vasileva, T.J. Morgan, An overview of the organic and inorganic phase composition of biomass, *Fuel,* vol. 94 (2012), pp. 1-33
[7] Barmina I, Lickrastina A, Zake M, Arshanitsa A, Solodovnik V and Telychian G 2013 *Latvian Journal of Physics and Technical Sciences* **50** N3 35-48
[8] Nordgren D, Hedman H, Padban N, Bostrom D and Ohman M 2013 *Fuel Processing Technology* **105** 52-58
[9] Wang C, Wu Y, Liu Q and Wang F 2011 *Fuel Processing Technology* **92** 5 1037-1041
[10] Barmina I, Valdmanis R and Zake M 2016 *Chemical Engineering Transactions* **52** 919.-924
[11] Zheng Y, Jensen P A, Jensen A D, Sander B and Junker H 2007 *Fuel* **86** (7-8) 1008-1020
[12] Shahzad K, Saleem M, Ghauri M, Akhtar J, Ali N and Akhtar A 2015 *Combustion Science and Technology* **187** (7) 1079-1092
[13] Thai S, Omar R, Kamal S M M and Biak D R A 2015 *Journal of Engineering Science and Technology* 97-109
[14] Ingole P M, Ranveer A C, Deshmukh S M and Deshmukh S K *International Journal of advanced Technology in Engineering and Science* **4** (6) 78-84
[15] Keshwani D R 2010 Microwave pretreatment of switchgrass for bioethanol production *PhD thesis* (University of York) p 258
[16] Lanigan B 2010 Microwave Processing of Lignocellulosic Biomass for Production of fuels *PhD thesis* (University of York) p 258
[17] Fang Z, Smith R L and Qi X 2014 Production of Biofuels and Chemicals with Microwave, *Sringer* p 271
[18] Barmina I, Līckrastiņa A, Valdmanis J, Valdmanis R, Zaķe M, Arshanitsa A, Solodovnik V and Telysheva G 2013 *Latvian Journal of Physics and Technical Sciences* **4** 34-48
[19] Arshanitsa A, Akishin Y, Zile E, Dizhbite T, Solodovnik V and Telysheva G 2016 *Renewable Energy* **91** 386-396
[20] Wang C, Wang F, Yang Q and Liang R 2009 *Biomass and Bioenergy* **33** 50-56
[21] Yang Q and Wu S 2009 *Cellulose Chemistry and Technology* **43** (4-6) 133-139
[22] Shen D K, Gu S, Luo K H, Bridgwater A V and Fang M X 2009 *Fuel* **88** 1024-1030
[23] Zhou H, Jensen A D, Glarborg P, Jensen P A and Kavaliauskas A 2005 *Fuel* **84** 389-403

Acknowledgments
The authors would like to acknowledge the financial support from the European Regional funding of the project Nr. 1.1.1.1/16/A/004

IOP Conf. Series: Materials Science and Engineering **355** (2018) 012019 doi:10.1088/1757-899X/355/1/012019

Computational Simulations of the Lateral-Photovoltage-Scanning-Method

S Kayser, A Lüdge and K Böttcher
Leibniz Institute for Crystal Growth, Max-Born-Straße 2, D-12489 Berlin, Germany

e-Mail: Stefan.Kayser@ikz-berlin.de

Abstract. The major task for the **L**ateral-**P**hotovoltage-**S**canning-Method is to detect doping striations and the shape of the solid-liquid-interface of an indirect semiconductor crystal. This method is sensitive to the gradient of the charge carrier density. Attempting to simulate the signal generation of the **LPS**-Method, we are using a three dimensional Finite Volume approach for solving the van Roosbroeck equations with COMSOL Multiphysics in a silicon sample. We show that the simulated LPS-voltage is directly proportional to the gradient of a given doping distribution, which is also the case for the measured LPS-voltage.

1. Introduction

Among the methods for crystal growth from the melt there are several techniques where the growing crystal is in permanent rotation inside of an inhomogeneous temperature field. For the Floating Zone technique this is, above all, caused by the gap in the heating coil. Consequently, each point of the melt sees a periodically changing temperature field leading to corresponding convection patterns. The additionally fluctuating doping concentration at the crystal-melt-interface leads to 3-dimensional concentration profiles roughly in form of a spiral staircase. The geometric shape of the 3D profiles follows the shape of the crystal-melt-interface. Making a cut through the crystal in growth direction results in a crystal surface with a periodical concentration pattern that reveals the growth history of the crystal-melt-interface, which again represents a finger print of the individual single crystal. The **L**ateral-**P**hotovoltage-**S**canning-Method (**LPS**) is known to be the method of choice for detecting these doping inhomogeneities in grown semiconductors, such as silicon [1], germanium [2] and silicon-germanium [3].

A first important question is, which local resolution can be achieved with respect to the mean free path of the charge carriers. In order to answer this question, a simulation of this measurement technique is established using a Finite Volume approach.

2. Experimental

Figure 1 shows the measurement setup of the LPS-Method. Here, a modulated laser ($f = 1$ kHz, square-wave) is focused ($\emptyset = (10 - 30)$ µm) on a semiconductor sample and generates excess photo carriers. If a dopant gradient exists at that observation point, the generated electrons and holes will drift in different directions due to the inner electric field [4]. Hence, a local dipole will be generated temporarily. This leads to a potential difference between the ohmic rim contacts (U_{LPS}), which can be amplified and detected with respect to the modulation frequency.

Content from this work may be used under the terms of the Creative Commons Attribution 3.0 licence. Any further distribution of this work must maintain attribution to the author(s) and the title of the work, journal citation and DOI.
Published under licence by IOP Publishing Ltd

The measurement is carried out by changing the focal point of the laser and detecting $U_{\mathrm{LPS}}(x, y)$. This signal is known to be proportional to the gradient of the dopant distribution: $U_{\mathrm{LPS}}(x, y) \propto \nabla_{\mathrm{x}} N_{\mathrm{D}}^{+}(x, y)$ [1].

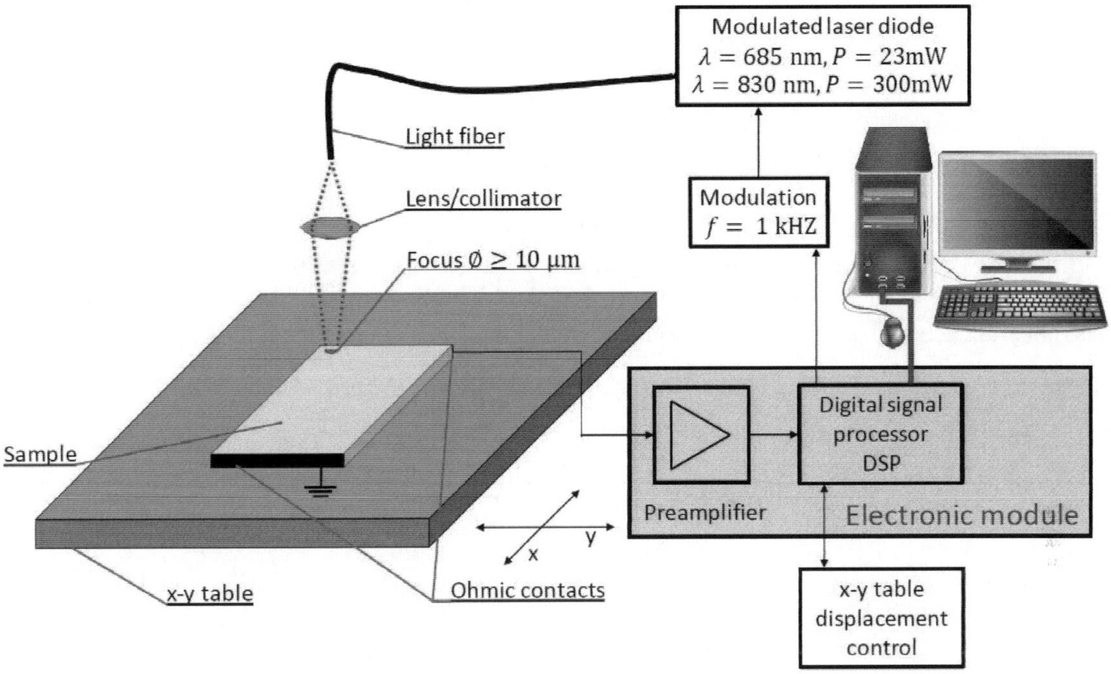

Figure 1. Scheme of the LPS-measurement setup. A modulated laser generates electrons/holes locally. If there is a gradient on the doping distribution at the observation point, a voltage difference will be detectable at the ohmic rim contacts. This voltage is amplified with respect to the modulation frequency.

3. Computational Simulation

3.1. Equation System

The fundamental equation system for the charge carrier flow and the electrostatic potential in a semiconductor structure is named after van Roosbroeck [5]. This system consists of three nonlinear coupled partial differential equations for the three physical variables, the electrostatic potential $\boldsymbol{\Psi}(\boldsymbol{r})$, the densities of electrons $n(\boldsymbol{r})$ and of holes $p(\boldsymbol{r})$. The potential and the charge carrier densities obey the Poisson equation of the electric field, which is in detail:

$$-\nabla \cdot \left(\epsilon_0 \epsilon_{\mathrm{r}} \nabla \boldsymbol{\Psi}(\boldsymbol{r})\right) = \mathrm{q}(p(\boldsymbol{r}) - n(\boldsymbol{r}) + N_{\mathrm{D}}^{+} - N_{\mathrm{A}}^{-}), \tag{1}$$

with the vacuum permittivity ϵ_0 and relative permittivity ϵ_{r}, the elementary charge q, and the densities of ionized donors and acceptors N_{D}^{+} and N_{A}^{-}.

The continuity equations for electrons (2) and holes (3) consider both the generation rate $G(\boldsymbol{r})$ and the recombination rate $R(\boldsymbol{r})$:

$$\frac{\partial n}{\partial t} - \frac{1}{\mathrm{q}} \nabla \cdot \boldsymbol{j}_{\mathrm{n}}(\boldsymbol{r}) = G(\boldsymbol{r}) - R(\boldsymbol{r}), \tag{2}$$

$$\frac{\partial p}{\partial t} + \frac{1}{\mathrm{q}} \nabla \cdot \boldsymbol{j}_{\mathrm{p}}(\boldsymbol{r}) = G(\boldsymbol{r}) - R(\boldsymbol{r}). \tag{3}$$

In the van Roosbroeck equation system (1), (2) and (3) the current densities are modelled as a sum of drift and diffusion currents:

$$-\boldsymbol{j}_{\mathrm{n}}(\boldsymbol{r}) = \mathrm{q}[\mu_{\mathrm{n}} n(\boldsymbol{r}) \nabla \boldsymbol{\Psi}(\boldsymbol{r}) - D_{\mathrm{n}} \nabla n(\boldsymbol{r})],$$
$$-\boldsymbol{j}_{\mathrm{p}}(\boldsymbol{r}) = \mathrm{q}[\mu_{\mathrm{p}} p(\boldsymbol{r}) \nabla \boldsymbol{\Psi}(\boldsymbol{r}) + D_{\mathrm{p}} \nabla p(\boldsymbol{r})].$$

3.2. Specifications

For the energy distribution of electrons and holes we choose the Fermi-Dirac statistics, hence, the diffusion coefficients of electrons D_{n} and holes D_{p}, are related to the mobilities of electrons μ_{n} and holes μ_{p} by the generalized Einstein relation

$$\text{electrons:} \quad \frac{D_{\mathrm{n}}}{\mu_{\mathrm{n}}} = \frac{\mathrm{k_B} T}{\mathrm{q}} g\left(\frac{n}{N_{\mathrm{c}}}\right),$$
$$\text{holes:} \quad \frac{D_{\mathrm{p}}}{\mu_{\mathrm{p}}} = \frac{\mathrm{k_B} T}{\mathrm{q}} g\left(\frac{p}{N_{\mathrm{v}}}\right),$$

where k_{B} is the Boltzmann constant, T is the temperature, and N_{c} and N_{v} are the conduction and valence band density of states, resp., g is a nonlinear factor ($g > 1$), which considers the Fermi-Dirac statistics.

The generation of electrons and holes are determined by the energy of the incoming laser beam. The laser is moved by a velocity v_{L} along the x-direction focussed on the surface of the silicon sample during time t. The laser beam intensity profile is assumed to have a Gaussian distribution, and the laser spot size is defined according to the traditional definition at which the diameter is taken transverse to the beam axis at which the intensity profile falls to $1/\mathrm{e}^2 \mathrel{\widehat{=}} 13.5\ \%$ of the intensity peak [6]. In terms of the Gaussian distribution, the standard deviation of the laser intensity in x and y-direction at the surface of the silicon sample is just a quarter of the $1/\mathrm{e}^2$-diameter, and we assume them to be identical $\sigma_{\mathrm{x}} = \sigma_{\mathrm{y}} = d_{1/\mathrm{e}^2}/4$ [6].

When denoting the peak power of the laser by P_0 and the photon energy by $E_{\mathrm{ph}}\,[\mathrm{Ws}] = \mathrm{hc}/\lambda$ with λ as wavelength of the laser beam, h as Planck constant, and c as light velocity, then the total photon flow rate is $I_{\mathrm{tot}} = P_0/E_{\mathrm{ph}} = P_0\,\lambda/\mathrm{hc}$, which still needs to be shaped by the two-dimensional Gaussian distribution which itself includes the movement of the laser beam in x-direction. Finally the photon flux is

$$I_{\mathrm{F}}(x,y,t)\left[\frac{1}{\mathrm{m^2 s}}\right] = I_{\mathrm{tot}} \frac{1}{2\pi\sigma_{\mathrm{x}}\sigma_{\mathrm{y}}} \exp\left[-\frac{(x - x_0 - v_{\mathrm{L}} t)^2}{2\sigma_{\mathrm{x}}^2} - \frac{(y - y_0)^2}{2\sigma_{\mathrm{y}}^2}\right]$$

The photon flux impinging at the surface penetrates into the depth of the crystal sample which is the (negative) z-direction. The penetration depth is practically limited due to the absorption coefficient $A_{\mathrm{c}}\,[1/\mathrm{m}]$. Nevertheless, there is a three-dimensional generation rate of electrons and holes:

$$G(x,y,z,t)\left[\frac{1}{\mathrm{m^3 s}}\right] = A_{\mathrm{c}} \exp[-A_{\mathrm{c}}|z|]\, I_{\mathrm{F}}(x,y,t).$$

We are denoting that the simulations are done for room temperature, neglecting the heating by the laser.

The recombination rate on the right side of equation (2) and (3) consists of the most common processes: the direct recombination R_{Direct}, the Shockley-Read-Hall recombination R_{SRH}, and the Auger recombination R_{Auger}:

$$R(x, y, z, t) = R_{\text{Direct}} + R_{\text{SRH}} + R_{\text{Auger}}.$$

Every recombination process is implemented in COMSOL Multiphysics and is realized by supplying material dependent recombination rates.

Figure 2 illustrates that all boundaries Γ_N of the platelet are electrically isolated except the end-faces of the silicon sample: one (Γ_1) of the two is set to the equilibrium potential, the other is established as an ideal ohmic contact with current connection ($i_{\Gamma_2}(t)$) to an external circuit. In order to detect the voltage change at the end-faces of the platelet, the experimental setup includes an internal resistance R in an electrical circuit model.

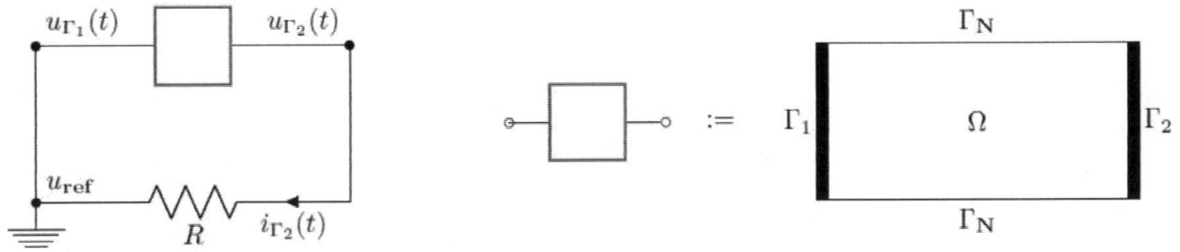

Figure 2. Scheme of the modelled electrical circuit and the boundary conditions of the sample Ω.

3.3. Numerical Procedure

The computational model was solved by using the software COMSOL Multiphysics V. 5.2a [7]. We use the "semiconductor module" and the "electrical circuit module".

The variables of the semiconductor part are the electric potential, the electron concentration and the hole concentration. The electric circuit part computes voltages and currents. Due to the thickness of the silicon platelet we have a 3-dimensional geometry which is meshed by a mapped grid ($x - y$-plane) that is swept in z-direction.

We held to the standard options of COMSOL for semiconductor problems: the Finite Volume Method is the numerical procedure, the numerical stability is reached by applying the Scharfetter-Gummel scheme [8].

4. Results and Discussion

The crystal is assumed to have a doping profile of boron or phosphorus, caused by segregation $N_{\text{D,seg}}$, which is characterized by a macroscopic gradient along the sample length and local fluctuations. This profile is modelled by the formula:

$$N_{\text{D}}^{+} = N_{\text{D,seg}} + N_{\text{D,fluct}} = N_{\text{D,0}}\left(1 + \frac{a}{L}x + b \cdot f(x, y, z)\right), \tag{4}$$

where $N_{\text{D,0}}$ is the background level of the dopant concentration, $L = 2.4$ mm is the sample length, x the length coordinate, a a control parameter for setting the macroscopic gradient of the dopant profile. The fluctuations of the doping concentration caused by temperature inhomogeneities are assumed as sinusoidal and suppressed by the prefactor $b = 10^{-5}$. The constant macroscopic gradient of the dopant profile, i.e. the linear behaviour of the profile, is assumed due to the small sample length: indeed there is a rather exponential distribution due to the segregation of, e.g., phosphorus in the melt during crystal growth.

The computation of the simulated LPS-voltage is shown in figure 3 for different values of the control parameter a in the time intervall $t \in [0.3\ 0.4]$ s. During this, the focus of the laser ($\sigma = 7.5$ μm) has moved between $x \in [0.9\ 1.2]$ mm (see upper axis). After a short loading phase of electrons and holes (where generation and recombination are not in equilibrium; see the vertical line at $t = 0.3$ s after the switch on) the computation yields a nearly constant LPS-voltage for each a-value when time runs.

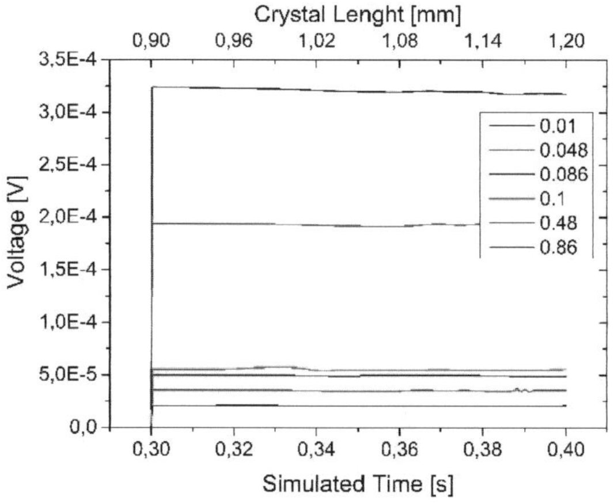

Figure 3. Result of time-dependent simulations of the LPS-voltage relative to the parameter a. Sample size: $th = 2\mu m$, $wd = 500\mu m$.

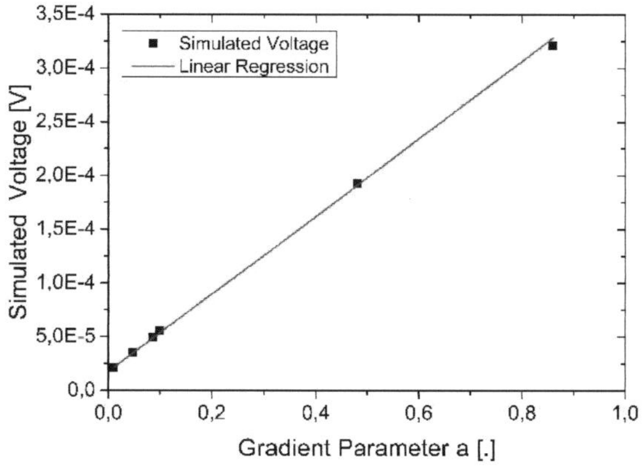

Figure 4. Dependency of the simulated LPS-voltages on the gradient parameter a.

Figure 4 shows the proportionality of the simulated LPS-voltage on the gradient of the dopant profile a. Additionally, Figure 4 reveals an offset of the LPS-voltage for $U_{\mathrm{Sim}}|_{a=0}$. In that case only the sin-shaped dopant fluctuations remain, hence, the offset cannot have physical reasons.

Therefore, we analysed, whether the offset depends on the model itself. We started to increase the sample width (from $wd = 150$ μm) and thickness (from $th = 2$ μm). As shown in figure 5, enlarging the sample width by a factor of 10 reduces the offset roughly by the same factor, however, the reduction of the offset is not a linear function of the sample width.

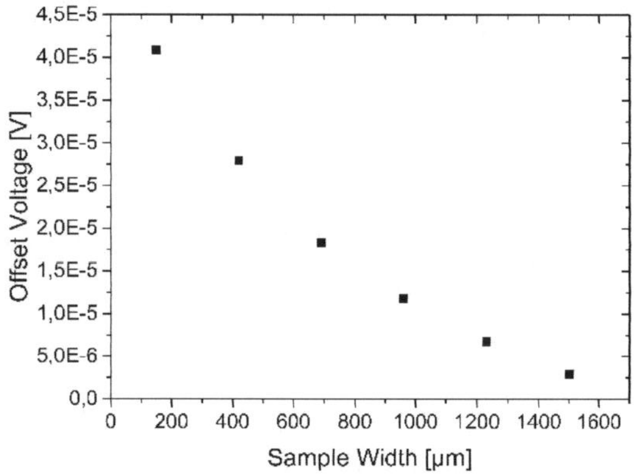

Figure 5. Behaviour of the offset with respect to the width of the simulated sample.

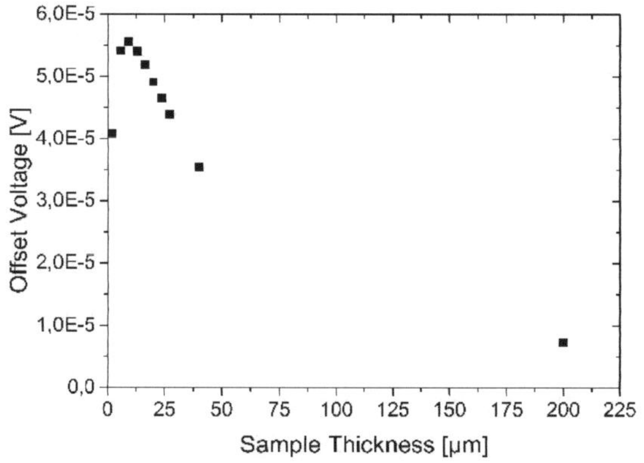

Figure 6. Dependency of the offset from the sample thickness. After an increase related to the penetration depth of the laser, a decreasing trend is visible.

The effect of the thickness on the offset is more complicated, see figure 6: as long as the sample thickness is below the penetration depth ($pd = 15.4$ µm), thickness enlargement increases the total number of charge carriers and the offset voltage is rising. But even before reaching the saturation of the total number of charge carriers, which may be reached at $th \approx 3pd$ (95 %), the offset decreases. The enlargement from the rather small thickness ($th = 2$µm) with maximum offset ($th_{\mathrm{max,off}} \approx 9\mu m$) to a thickness of $th = 200$ µm reduces the offset by a factor of eight. Because of $th \gg pd$, the non-reflected energy of the laser is absorbed (99.9998 %) by the semiconductor and transmission is highly suppressed. Combining the increased thickness and width of the sample, the offset is reduced tremendously and, thus, is negligible.

The linear dependency of the simulated LPS-voltage on a and the knowledge that $U_{\mathrm{LPS}} \propto \nabla_{\mathrm{x}} N_{\mathrm{D}}^{+}$ [1] allows the following relation:

$$U_{\mathrm{Sim}} \propto a \propto \nabla_{\mathrm{x}} N_{\mathrm{D}}^{+} \propto U_{\mathrm{LPS}}.$$

This shows that the simulated LPS-voltage corresponds to the measured LPS-voltage.

5. Conclusions and Outlook

For large enough samples ($wd \geq 1500$ µm, $th \geq 200$ µm) we have shown that the simulated LPS-voltage is directly proportional to the measured voltage $U_{\mathrm{Sim}} \propto U_{\mathrm{LPS}}$ as expected from experimental results. Simulation costs are high for such large samples, but can be reduced using symmetry assumptions (e.g. $N_{\mathrm{D}}^{+}(x, y) = N_{\mathrm{D}}^{+}(x, -y)$).

Now the local resolution can be calculated and related to the LPS-measurement method. Also, for a given mean charge carrier density (e.g. via Hall-measurements) it is possible to calculate the absolute value of the doping concentration by integration of the LPS-measurement results. This could lead to a resistivity profile with an improved local resolution compared to 4-point-measurement systems. We estimate the order of magnitude of the resolution to be hundred microns.

References

[1] Lüdge A, Riemann H, 1997 Doping Inhomogeneities in Silicon Crystals Detected by the Lateral Photovoltage Scanning (LPS) Method., *Inst. Phys. Conf.* Ser. No **160,** 145ff

[2] Wünscher M, Lüdge A, Riemann H, 2011 Crucible-free pulling of germanium crystals., *Journal of Crystal Growth,* Vol. **318**, pp 1039-1042

[3] Abrosimov N V, Lüdge A, Riemann H, Schröder W, 2002 Lateral photovoltage scanning (LPS) method for the visualization of the solid-liquid interface of $Si_{1-x}Ge_x$ single crystals, *Journal of Crystal Growth*, Vol. **237-239**, pp 356-360

[4] Schurig T, Matz H, Drung D, Lüdge A, Riemann H, 1997 Nondestructive Wafer Inspection utilizing SQUIDs, *Inst. Phys. Conf. Ser.* No **160**, 149ff

[5] van Roosbroeck W, 1950 Theory of the flow of electrons and holes in germanium and other semiconductors, *Bell Syst. Tech. J.*, Vol. **29**, pp 560-607

[6] Marshall G F, Stutz G E (Eds.), 2012 Handbook of Optical and Laser Scanning, *CRC Press*, Boca Raton

[7] COMSOL Multiphysics: Semiconductor Module, Users Guide V. 5.2a, 2016, *COMSOL AB*

[8] Scharfetter D L, Gummel H K, 1969 Large signal analysis of a silicon Read diode, *IEEE Trans. Electron. Dev.*, Vol. **16**, 6477

IOP Conf. Series: Materials Science and Engineering **355** (2018) 012020 doi:10.1088/1757-899X/355/1/012020

Numerical modelling of surface waves generated by low frequency electromagnetic field for silicon refinement process

V Geža[1], J Venčels, Ģ Zāģeris and S Pavlovs

University of Latvia, LV-1002, Riga, Latvia

[1]E-mail: vadims.geza@lu.lv

Abstract. One of the most perspective methods to produce SoG-Si is refinement via metallurgical route. The most critical part of this route is refinement from boron and phosphorus, therefore, approach under development will address this problem. An approach of creating surface waves on silicon melt's surface is proposed in order to enlarge its area and accelerate removal of boron via chemical reactions and evaporation of phosphorus.

A two dimensional numerical model is created which include coupling of electromagnetic and fluid dynamic simulations with free surface dynamics. First results show behaviour similar to experimental results from literature.

1. Introduction

The exhaustion of fossil fuel resources, global warming, and increased energy consumption all make finding alternative energy sources important. In terms of environmental impact, solar energy is one of the most perspective and attractive sources. However, development of solar power is limited by the solar power costs in comparison to other power sources. As shown in [1], by 2020 the energy cost for solar power is still expected to be higher than for hydro, onshore wind, geothermal, nuclear and even biomass energy sources. Reduction of solar cell production costs is necessary for this technology to develop, and one solution is to reduce the cost of raw materials like polysilicon.

In 2013, production of polysilicon constituted approx. 23% of the total solar cell costs [2]. This number can be reduced by lowering energy consumption during solar grade silicon (SoG-Si) production. One of the most efficient ways to produce SoG-Si is the metallurgical route of refining. It consumes about 30 *kWh/kgSi* (kilowatt-hours per kilogram of silicon) of energy. In comparison, chemical route such as the Siemens process, which is the most widely used refining process nowadays, consumes more than 70 *kWh/kgSi* [3]. In 2013, more than 200 000 *tons* of polysilicon was produced worldwide, which could result in more than an 8 *TWh* reduction if all of the silicon was produced via the metallurgical route rather than with the Siemens process.

Currently, the polysilicon market is recovering from crisis and new technologies that ensure reduction of production costs can give a significant advantage in a competitive SoG-Si market. For this purpose sophisticated simulation tools are required, due to complexity of underlying phenomena, and multiphysical nature of this process, which includes electromagnetics, thermal physics and fluid dynamics. Free surface phenomena is of particular interest, because of its complexity and huge computational requirements due to necessity of coupling between different simulation tools or packages. There are several papers on free surface dynamics in electromagnetically driven flows published in recent years. In [4] it is shown that volume of fluid (VOF) method gives good agreement with experimental results. Drawback of use of VOF for problems for very dynamic surfaces is large

Content from this work may be used under the terms of the Creative Commons Attribution 3.0 licence. Any further distribution of this work must maintain attribution to the author(s) and the title of the work, journal citation and DOI.
Published under licence by IOP Publishing Ltd

IOP Conf. Series: Materials Science and Engineering **355** (2018) 012020 doi:10.1088/1757-899X/355/1/012020

computational time. On the other hand, when surface deflection is not large, VOF allow calculation of free surface dynamics in reasonable time [5]. In all mentioned works, it is common assumption that EM frequency is too large to influence surface dynamics at its frequency. However, in problem under consideration in this work, it is not the case. In this work low frequency EM field is used and time dependence of EM variables has to be taken into account. The main task of this work is creation of numerical tools for simulations of silicon refinement from boron and phosphorus with use of advanced electromagnetic impact.

2. Potential for EM technologies in silicon refinement

In many investigations phosphorus removal is found to be the most challenging task, since the purity threshold (< 1 *ppm*) was not reached [6]. Although, oxidation and slag refinement is simple process for removing phosphorus from silicon, it is insufficient to reach low purity threshold. Mostly used approach for phosphorus removal is evacuation, it is dependent on free surface area and bulk melt stirring. Refinement enhancement can be achieved by superposing DC and AC magnetic fields [7]. Furthermore, this impact creates also mixing effect in the bulk melt, which is also important for boron removal. In comparison, classical induction stirring can enhance only bulk melt mixing, leaving surface area size unchanged.

Boron removal by oxidizing bubbles was already shown in [8] where significant splashing of silicon was observed, as a result blowing of oxidizing gas on free surface of the silicon was preferred instead. Also, no attempt was made to adapt electromagnetic stirring for bubble flow control, which was shown to be possible in [9]. Furthermore, high velocity shear in melt can lead to gas bubble breakup thus enlarging gas-liquid interface area and increasing boron removal rate.

Other possibility for boron removal could be blowing of oxidizing gas on free surface, which is already well known [10], and with additional creation of surface waves by means of electromagnetic impact thus creating capillary waves and increasing surface area at least twice. Previously mentioned electromagnetic impact technology (superposition of DC and AC fields) can be used for boron evaporation, because it enhances surface area and stirring.

Highest boron refining rates can be achieved with plasma refining, but it has high energy demand. In reactive gas refining, concentration of boron changes according to first order rate law

$$ln\left[\frac{C_{B,0}}{C_{B,t}}\right] = k_B\frac{A}{V}t \qquad (1)$$

Here C_B – boron concentration with indices after comma indicating time, A is the area of interfacial surface between oxidizing gas and silicon melt, V is the volume of silicon melt, t is time, k_B is boron mass transfer coefficient at the surface, measured in $\mu m/s$. k_B is determined experimentally and is dependent on oxidizing gas supply rate, type of oxidizing gas and melt temperature. Creation of surface waves would increase A/V ratio, thus speeding up boron removal. Table 1 summarizes k_B from literature with different oxidizing gases, gas flow rates and temperatures.

Data from Table 1 shows that the most important factor for boron removal is oxidizing gas and its flow rate.

According to [13], the same equation (1) is applicable for phosphorus removal, literature data is also shown in Table 1. For phosphorus, temperature is a more important factor. Another important factor is pressure. According to [14], the phosphorus mass transfer coefficient on the surface of the silicon melt strongly depends on pressure. Above 1 *Pa* pressure, the transfer rate of phosphorus declines very rapidly. On the other hand, phosphorus mass transfer coefficient reaches asymptotical value below 0.1 *Pa* and therefore further decrease in pressure would not be beneficial.

For removal of both boron and phosphorus removal, surface waves can be used to enlarge surface area and increase refinement rate. However, it is not clear how large the surface enlargement will be and how it will influence boron diffusion in gas boundary layer. For this purpose, the first step is creation of a numerical model which can simulate melt's surface waves induced by electromagnetic impact.

Table 1. Silicon refining from boron and phosphorus summary from literature.

Oxidizing gas	Melt temperature T, °C	Flow rate, Nl/min	k_B, μm/s	Reference
Ar – 97 %, H_2O – 3%	1450	3	7.0	[11]
N_2 – 97 %, H_2O– 3%	1450	3	6.0	
H_2 – 97 %, H_2O– 3%	1450	3	13.0	
H_2 – 24.2 %, Ar – 72.6 %, H_2O– 3.2%	1500	3	4.3	[12]
H_2 – 48.4 %, Ar – 48.4 %, H_2O– 3.2%	1500	3	8.6	
H_2 – 72.6 %, Ar – 24.2 %, H_2O– 3.2%	1500	3	9.2	
H_2 – 97 %, H_2O– 3%	1450	3	13.0	[10]
H_2 – 97 %, H_2O– 3%	1500	2	9.0	
H_2 – 97 %, H_2O– 3%	1500	14	44.6	
Boron / Phosphorus	Melt temperature T, °C	Pressure, Pa	k_P, μm/s	Reference
	1500	0.5	2.28	[13]
	1600	0.5	4.93	

3. Simulation model

A 2D axisymmetric problem of mercury with free surface is considered. The mercury tank is surrounded by inductor, which creates alternating magnetic field, see Figure 1 (a) for illustration. A similar device and experiment was described in [15]. This mercury experiment is only used for model validation purposes.

In this model, mercury properties were set in the melt domain – electrical conductivity σ_{melt}=1.04·10^6 *S/m*, kinematic viscosity ν_{melt} = 1.14·10^{-7} *m²/s*, density ρ = 13534 *kg/m³*, surface tension γ = 0.5 *N/m* and contact angle θ = 90°. The inductor has homogeneous current distribution with density J_{ampl} = 2.5·10^6 *A/m²* and frequency f = 14 *Hz*. Turbulence model *k-ε* is used, and maximal Courant number is set to 0.4. The transient simulation reached flow time 4 *sec*.

For numerical simulations we use two open-source software packages – the finite element multi-physics simulation code Elmer and computational fluid dynamics code *OpenFOAM*. Efficient parallel coupling between both codes is done using *EOF-Library* [16].

Free surface is computed using Volume of Fluid (VOF) method, which is implemented in interFoam solver in *OpenFOAM*. Euler scheme is used for time discretization, linear schemes for spatial discretization of all other terms in momentum equations. Equations for turbulent properties k and ε are discretized using upwind scheme.

Computational mesh for *OpenFOAM* has 51k cells, mesh for *Elmer* has 46k elements. We used very fine mesh with cell size 0.085x0.35mm in the region where surface can be present during simulation.

IOP Conf. Series: Materials Science and Engineering **355** (2018) 012020 doi:10.1088/1757-899X/355/1/012020

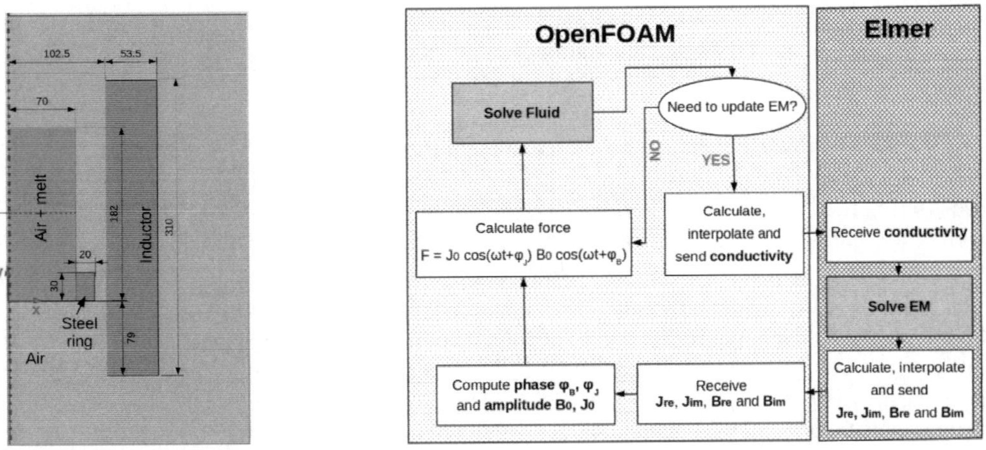

(a) (b)

Figure 1. (a) Schematic drawing for 2D axisymmetric model, where vertical dashed line is symmetry axis and horizontal dashed line is melt's initial filling level. x is radial coordinate, y is axial coordinate, z is azimuthal coordinate (b) Computational scheme for coupled *OpenFOAM* and *Elmer* simulation.

Computation scheme is shown in Figure 1 (b). *OpenFOAM* package solves fluid dynamics with free surface using volume of fluid method, then computes electrical conductivity of gas-and-melt mixture and sends it to Elmer package. Elmer solves the complex problem for the time-harmonic electromagnetic (EM) field and sends back to *OpenFOAM* the complex components of electrical current density and magnetic field density.

Every time the EM field is updated, the amplitudes are computed

$$J_0^i = \sqrt{J_{\Im}^2 + J_{\Re}^2}$$
$$B_0^i = \sqrt{B_{\Im}^2 + B_{\Re}^2} \quad (2)$$

and the corresponding phases

$$\varphi_J^i = atan2\left(J_{\Im}^i, J_{\Re}^i\right)$$
$$\varphi_B^i = atan2\left(B_{\Im}^i, B_{\Re}^i\right) \quad (3)$$

for current and magnetic field densities, where i represents x, y and z components (radial axial and azimuthal respectively), and *atan2* is the arctangent function with two arguments that gives the angle in radians between the x-axis and the specified point.

Then, at every *OpenFOAM* timestep the *Lorentz* force is updated

$$F^x = exp\left(-\frac{2}{\omega t}\right) \cdot J_0^z cos\left(\varphi_J^z + \omega t\right) \cdot B_0^y cos\left(\varphi_J^y + \omega t\right) + F_{corr}^x$$
$$F^y = -exp\left(-\frac{2}{\omega t}\right) \cdot J_0^z cos\left(\varphi_J^z + \omega t\right) \cdot B_0^x cos\left(\varphi_J^x + \omega t\right) + F_{corr}^x \quad (4)$$
$$F^z = 0$$

where ω is angular frequency and F_{corr} is a force correction due to interaction of velocity and magnetic field:

$$F_{corr}^x = \sigma \cdot B_0^y cos\left(\varphi_J^y + \omega t\right) \cdot \left(B_0^x cos\left(\varphi_B^x + \omega t\right) \cdot U_y(t) - B_0^y cos\left(\varphi_B^y + \omega t\right) \cdot U_x(t)\right)$$
$$F_{corr}^y = \sigma \cdot B_0^x cos\left(\varphi_J^x + \omega t\right) \cdot \left(B_0^y cos\left(\varphi_B^y + \omega t\right) \cdot U_x(t) - B_0^x cos\left(\varphi_B^x + \omega t\right) \cdot U_y(t)\right)$$

An exponential multiplier in equation (4) provides a smooth increase of *Lorentz* force during the first few periods and reduces unwanted splashes at the beginning of simulation.

4. Results

Turbulent flow ($Re \approx 2.5 \cdot 10^4$) inside mercury vessel has two toroidal vortices similar to a classical induction crucible. Steel ring near the bottom of the crucible (see Figure 1 (a)) significantly changes the size ratio of two vortices – the upper vortex becomes larger and more dominant due to field screening near bottom of crucible. Figure 2 (a) shows instantaneous velocity field in the melt. The velocity vectors in this figure that have also vertical component only indicate that melt surface is moving and there is not net melt mass moving through the interface.

Slow change of the EM force in the melt causes creation of surface waves near the outer wall, and subsequent propagation of waves radially inward. Waves have very small amplitude (below 0.5 mm) and they are hardly distinguishable in Figure 2(a)

Frequency analysis of local melt height in three points ($x = 0$ on axis, $x = 35$ *mm*, and $x = 70$ *mm* near wall) is shown in Figure 2 (b). There is one very strong peak at 28 *Hz* that is connected with the frequency of the EM force, which is double the EM field frequency. Analysis of surface data shows that there are 10 full waves on the surface, which gives wavelength $\lambda \approx 7$ mm, same result as in [15].

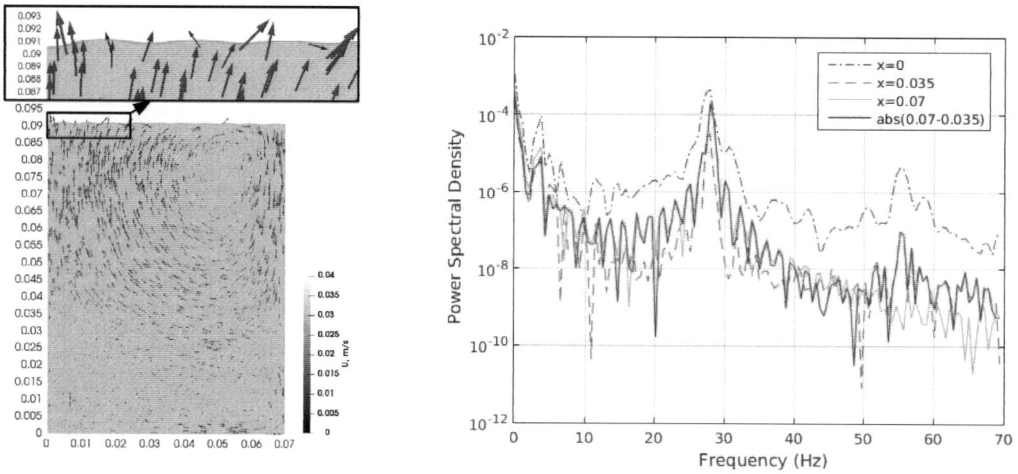

(a) (b)

Figure 2. (a) Melt's surface shape and velocity distribution at 1.5 *s*. Small waves can be visible in magnification above (b) Wave frequencies at three different radial coordinates.

The time dependence for the surface shape of melt during one 14 Hz period can be seen in Figure 3 (a). It was observed that the surface has regular structure with waves traveling radially inward. These data also allow estimation of wave phase speed by means autocorrelation of point data. Obtained result is $c \approx 19.5$ cm/s, and it corresponds well to $\lambda/T = 19.6$ cm/s.

Another comparison was done with experimental velocity measurement data from [15]. Figure 3 (b) shows radial velocity component plot on vertical line r = 49 mm. There is a good agreement between experimental and simulation data with ring. Initially simulations were done without steel ring, which was not mentioned in [15], but was found in other work of the same author [17]. This result shows how strongly steel ring changes the flow structure.

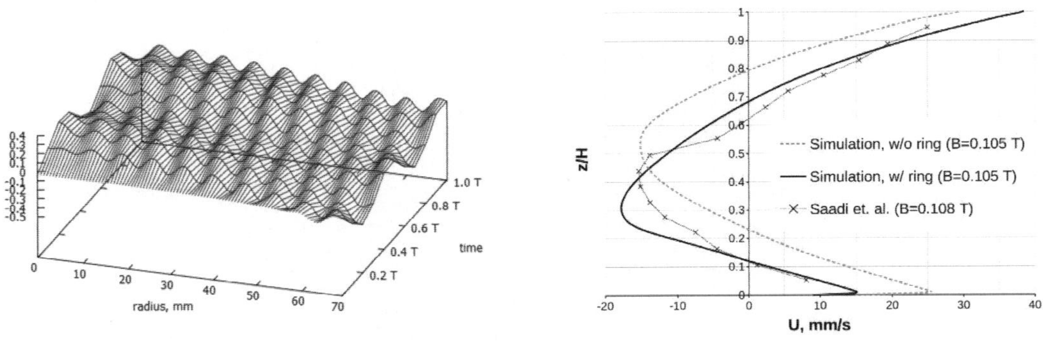

(a) (b)

Figure 3. (a) Melt surface height radial dependence in time, B=0.108 T. Vertical axis showing melt height in millimetres. Radius is x coordinate. T is period of EM field, T=1/14 Hz = 0.071 s (b) Comparison of simulation and experimental data for velocity on vertical line r=49 mm.

There is uncertainty about wave height in experimental data [15] and therefore validation of model requires further step. Furthermore, obtained results show good agreement, but obtained wave heights are too small for practical applications of silicon refinement. For this purpose, the magnetic field on crucible axis was increased from 0.108 T to 0.257 T.

For this case, wave structure differ significantly (Figure 4). Wave amplitude is much larger (up to 2 mm), but there is no regular structure of waves moving toward centre. Waves here can be separated in three regions.

Most outer region, between radius values 45 and 70 mm, is characterized by waves traveling radially inward. However, they differ from small wave case, because here one wave is created for every two force pulses. In small wave case one waves was created each force pulse.

In middle region (15-45 mm) waves are travelling radially toward the centre initially and when waves reach the centre, waves travelling away from centre appear. After two seconds, first standing waves start to appear when oppositely travelling waves interfere.

In innermost region spatial waves are almost not distinguishable, but whole region oscillates in time.

It is visible that for high magnetic field case, wave height strongly decrease in radial direction – there is approx. 2 mm amplitude near outer wall and about 5 times smaller amplitude near the centre of crucible.

This finding with higher amplitude waves correspond to experimental findings in [15] where at high magnetic field, structure of waves becomes non-axisymmetric. For this reason, extension of this work with 3D simulations is planned.

5. Conclusions and outlook

Developed numerical model showed that waves are formed on melt's surface with low frequency EM field excitation. Results are in good agreement with literature data. However, these waves are too small to be used in practice.

It was shown that waves with larger amplitude differ significantly in structure and therefore require more thorough study and experimental validation. There are very few references on numerical modelling of such problems and therefore these promising results are important step towards better understanding and capturing of EM induced waves on liquid metal surface.

Acknowledgments

This work was funded by *European Regional Development Fund* under contract "*Refinement of metallurgical grade silicon using smart refinement technologies*" (No. 1.1.1.1/16/A/097).

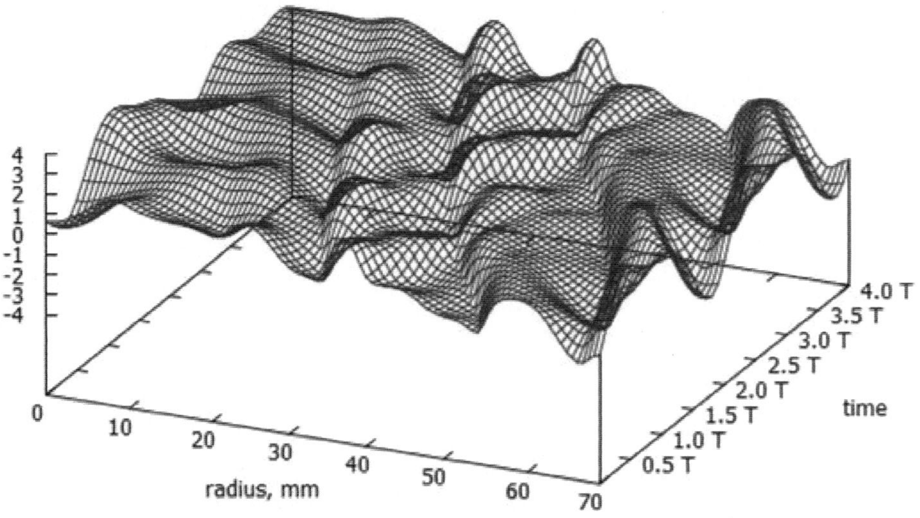

Figure 4. Melt surface height radial dependence in time, B=0.257 T.

References

[1] U.S. Energy Information Administration (EIA) – Source // www.eia.gov (retrieved 2016-06-07)

[2] GTM Research // www.greentechmedia.com/articles/read/solar-cost-reduction-drivers-in-2017 (retrieved 2016-06-06)

[3] Fu R, James T L, Woodhouse M 2015 *IEEE Journal of Photovoltaics* **5** 515

[4] Spitans S, Baake E, Jakovics A, Franz H 2017 *Int. J. Appl. Electromagn. Mech* **53** S61.

[5] Beckstein P, Galindo V, Schonecker A, Gerbeth G 2017 *IOP Conf. Ser.: Mater. Sci. Eng.* **228** 012018

[6] Bojarevičs A, Beinerts T, Grants I, Kaldre I, Šivars A, Gerbeth G, Gelfgat Yu 2015 *Magnetohydrodynamics* **51** 437

[7] Khattak C P, Joyce D B, Schmidt F 2002 *Solar Energy Materials and Solar Cells* **74** 77

[8] Sortland Ø S 2015 *Boron removal from silicon by steam and hydrogen. PhD Thesis* (Trondheim: Norwegian University of Science and Technology) p 268

[9] Sand U, Yang H, Eriksson J-E, Bel Fdhila R 2009 *Steel Research International* **80** 441

[10] Sortland Ø S, Tangstad M 2014 *Metallurgical and Materials Transactions* E **1** 211

[11] Safarian J, Thang K, Hildal K, Tranell G 2014 *Metallurgical and Materials Transactions* E **1** 41

[12] Nordstrand E F, Tangstad M 2012 *Metallurgical Material Transactions* B **43** 814

[13] Safarian J, Tangstad M 2012 *High Temperature Material Processing* **31** 73

[14] Zheng S, Safarian J, Seongho S, Sungwook K, Tangstad M, Luo X 2011 *Transactions of nonferrous metals society of China. Metals*, **21** 697

[15] Saadi B, Bojarevics A, Fautrelle Y, Etay J 2005 *Récents Progrès en Génie des Procédés, Numéro* 92 (Paris) (ISBN 2-910239-66-7)

[16] Vencels J, Jakovics A, Geza V, Scepanskis M 2017 *XVIII Int. UIE-Congress* (Hannover) (Essen Vulcan-Verlag GmbH) p 312

[17] Bachir Saadi. Contrôle électromagnétique des transferts de masse aux interfaces liquide/liquide. Electromagnétisme. INSTITUT NATIONAL POLYTECHNIQUE DE GRENOBLE, 2006. France.

IOP Conf. Series: Materials Science and Engineering **355** (2018) 012021 doi:10.1088/1757-899X/355/1/012021

Development of an electromechanical principle for wet and dry milling

Bernd Halbedel and Oleg Kazak

Department of Mechanical Engineering, Group for Inorganic-Nonmetallic Materials, Technische Universität Ilmenau, Gustav-Kirchhoff-Straße 6, D-98684 Ilmenau, Germany

E-mail: bernd.halbedel@tu-ilmenau.de

Abstract. The paper presents a novel electromechanical principle for wet and dry milling of different materials, in which the milling beads are moved under a time- and local-variable magnetic field. A possibility to optimize the milling process in such a milling machine by simulation of the vector gradient distribution of the electromagnetic field in the process room is presented. The mathematical model and simulation methods based on standard software packages are worked out. The results of numerical simulations and experimental measurements of the electromagnetic field in the working chamber of a developed and manufactured laboratory plant correlate well with each other. Using the obtained operating parameters, dry milling experiments with crushed cement clinker and wet milling experiments of organic agents in the laboratory plant are performed and the results are discussed here.

1. Introduction

Milling is an important but complicated and energy consuming process in the production of materials. The interest in improvement of milling processes in numerous industrial applications such as finest milling of pharmaceutical organic agents [1], raw materials for ceramics, material for building industries [2] as well the recycling [3] has increased steadily. The novel electromechanical principle can be a solution for improvement of milling in numerous applications, but still need methods and tools for the design and optimization of the milling machine.

A laboratory plant that we have developed and manufactured previously has shown very good milling results [4]. Furthermore, the power consumption is very low as a result of direct power supplying to the grinding media. The biggest energy fraction, around 74 % of the total power supplied to the machine, is going to the milling process itself and only 23 % to ohmic losses as well as 3 % to iron losses in the exciter systems [4].

The main running parameter of such a machine for milling (a so-called EMZ plant) is the magnetic flux density and the resulting electromagnetic force distribution in the process room. For future development of this principle and design of new machines it is very important to develop a suitable model for numerical simulation of the resulting electromagnetic force distribution and its optimization. The model for numerical simulation is developed based on ANSYS Maxwell. The convergence of the calculation mesh is tested with different software setup as well as with different applied currents and current frequencies. Simulations are carried out in 2D and 3D. Verification is done by comparing results in ANSYS Maxwell with COMSOL Multiphysics as well with experimental measurements of the electromagnetic field in the process room of the laboratory EMZ installation. Dry milling experiments of crushing cement clinker and wet milling experiments of organic agents in the laboratory plant are

Content from this work may be used under the terms of the Creative Commons Attribution 3.0 licence. Any further distribution of this work must maintain attribution to the author(s) and the title of the work, journal citation and DOI.
Published under licence by IOP Publishing Ltd

IOP Conf. Series: Materials Science and Engineering **355** (2018) 012021 doi:10.1088/1757-899X/355/1/012021

executed using the obtained operating parameters. The results of these milling experiments are discussed.

2. Electromechanical milling principle (EMZ)

The electromechanical milling principle is based on the moving of hard magnetic milling beads under the influence of the electromagnetic force $\vec{F}_{\nabla\vec{B}}$ generated by interaction between the vector gradient $\nabla\vec{B}$ of the magnetic field $\vec{B}$ and the fixed magnetization $\vec{M}$ of the milling beads as shown in Figure 1.

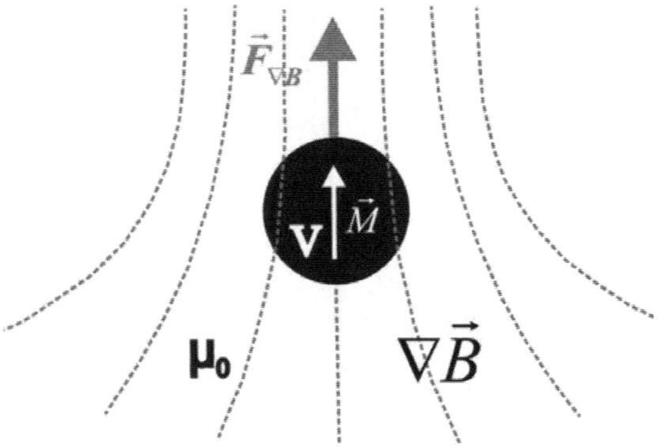

Figure 1. Magnetic force $\vec{F}_{\nabla\vec{B}}$ on a milling bead with a magnetization $\vec{M}$ and the volume V in a magnetic field $\vec{B}$ with a vector gradient $\nabla\vec{B}$.

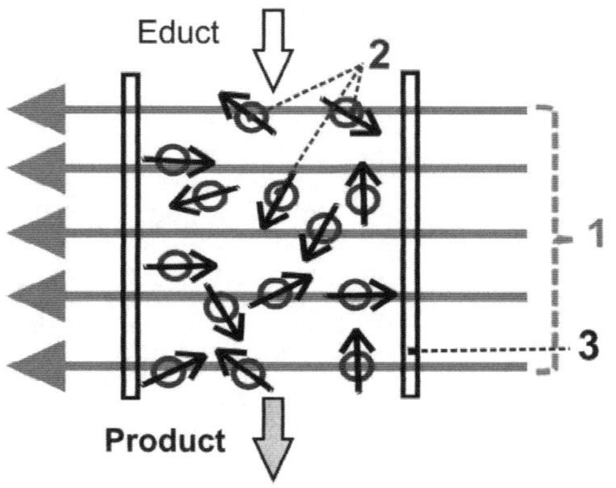

Figure 2. Principle of electromechanical milling (EMZ)
1 – time- and local-variable magnet field, 2 – hard magnetic milling beads, 3 – process room.

The generated force of one milling bead is given by equation (1):

$$\vec{F}_{\nabla\vec{B}} = (\vec{p}_M \cdot \nabla)\vec{B}, \tag{1}$$

where $\vec{p}_M$ is the vector of the magnetic moment, given for milling beads by

$$\vec{p}_M = \int_V \vec{M} dV = \frac{1}{\mu_0}\int_V \vec{J} dV, \tag{2}$$

with $\vec{M}$ – the fixed permanent magnetization of the milling beads, $\vec{J}$ – its magnetic polarization, V – the volume of a milling bead and $\mu_0 = 4\pi \cdot 10^{-7}$ Vs/Am – the absolute permeability.

With the simplifying assumptions that the magnetization of the milling bead is $\vec{M} = M = const.$ in its volume and directed in the same direction as $\nabla\vec{B}$ we can deduce equation (3) for description of the electromagnetic force density $\vec{f}_{\nabla\vec{B}}$ on one milling bead:

$$\vec{f}_{\nabla\vec{B}} = \frac{\vec{F}_{\nabla\vec{B}}}{V} = M \cdot (\nabla\vec{B}) = \frac{1}{\mu_0} J \cdot (\nabla\vec{B}), \tag{3}$$

with M – magnitude of $\vec{M}$ and J – magnitude of $\vec{J}$.

As revealed in equation (3) the vector gradient of the magnetic field $\nabla\vec{B}$ dictates the direction of the force density $\vec{f}_{\nabla\vec{B}}$ and the product $J \cdot (\nabla\vec{B})$ determines their magnitude.

Additionally, if the magnetic vector gradient is time- and local-dependent, then a time- and local-dependent force density distribution $\vec{f}_{\nabla\vec{B}}(r,\varphi,z,t)$ is generated in the process room, consequently an intensive relative movements between several milling beads are induced with different types of mechanical stresses so that the novel electromechanical milling principle can be used for disintegration of biomass, dry or wet grinding of raw materials and autogenously grinding of ferrites [4, 5].

The basic scheme of the electromechanical milling laboratory plant EMZ-LAI developed and manufactured previously by us is presented in Figure 2.

The process room is an annular gap bordered with non-ferromagnetic materials and surrounded internally and externally with a rotationally symmetrical exciter systems, which generate a rotating magnetic field $\vec{B}(r,\varphi,z,t)$. The designs of the exciter systems are analogical to an AC motor but with a large air gap between the outer (ESI) and the inner (ESII) exciter systems (cf. figure 3), which is used for milling.

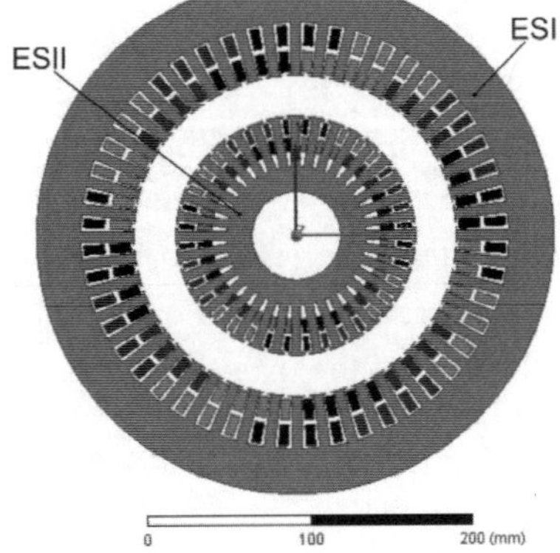

Figure 3. 2D cross section of the exciter systems ESI and ESII with the coils distribution, number of phases m=3, number of poles 2p=4.

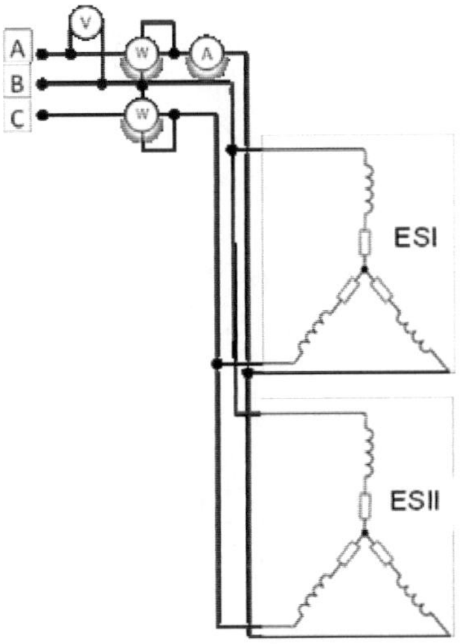

Figure 4. Electric connections of the windings from ESI and ESII to symmetric 3-phase system A, B and C with metering elements to measure the active power, voltage and current.

The windings of the exciter systems are performed with three phases and four poles. Figure 4 shows the electric connections of the windings from ESI and ESII to symmetric 3-phase system A, B, C.

3. Model for calculation of the electromagnetic design

The electromagnetic design of the EMZ machine is based on the consideration of equation (3) and the following important steps:

a) determination of the required vector gradient $\nabla\vec{B}$ of magnetic field distribution $\vec{B}(r,\varphi,z,t)$ in the air gap with an existing EMZ plant and
b) determination of a new design with the lowest possible losses at the required vector gradient $\nabla\vec{B}$.

For the determination of the vector gradient distribution $\nabla\vec{B}(r,\varphi,z,t)$, it is necessary to simulate the magnetic field distribution $\vec{B}(r,\varphi,z,t)$ in the air gap by equation (4)

$$\vec{B} = rot\vec{A} \qquad and \qquad div\vec{B} = 0\,, \tag{4}$$

where $\vec{A}$ is the magnetic vector potential resulting from the design of the windings and electric current distribution in the exciter systems.

Then $\nabla\vec{B}(r,\varphi,z,t)$ is calculable using equation (5):

$$\nabla \vec{B} = \left(\frac{\partial B_r}{\partial r} + \frac{1}{r}\frac{\partial B_r}{\partial \varphi} + \frac{\partial B_r}{\partial z} - \frac{B_\varphi}{r} \right) \vec{e}_r + \left(\frac{\partial B_\varphi}{\partial r} + \frac{1}{r}\frac{\partial B_\varphi}{\partial \varphi} + \frac{\partial B_\varphi}{\partial z} + \frac{B_r}{r} \right) \vec{e}_\varphi + \left(\frac{\partial B_z}{\partial r} + \frac{1}{r}\frac{\partial B_z}{\partial \varphi} + \frac{\partial B_z}{\partial z} \right) \vec{e}_z \quad (5)$$

The losses in the exciter systems of an EMZ plant are calculated as sum of core losses $P_{v,U}$ in its laminated cores and ohmic losses $P_{v,R}$ in the coils

$$P_v = P_{v,R} + P_{v,U} \quad (6)$$

with

$$P_{v,R} = \frac{1}{2\sigma}\int_{V_c} \vec{j}^2 dV \quad (7)$$

where σ is the electrical conductivity of the coils, $\vec{j}$ is the current density distribution in the windings and V_c is the winding volume,

and

$$P_{v,U} = \int_{V_{lc}} \left[K_h \cdot f \cdot \hat{B}^2 + K_c \cdot (f \cdot \hat{B})^2 + K_e \cdot (f \cdot \hat{B})^{1.5} \right] dV \,, \quad (8)$$

where f is the frequency of the electric currents in the coils of the windings, $\hat{B}$ are the values of the peak magnetic flux density in the laminated cores of the exciter systems, V_{lc} is the volume of the laminated cores, K_h is the hysteresis loss coefficient, K_c is the eddy current loss coefficient and K_e is the excessive loss coefficient. These coefficients are known from the data of the iron sheets producer.

To build a model for simulation of the electromagnetic field in ANSYS Maxwell, it is necessary to include the design of the laminated cores and the coils distribution of the inner and outer systems of the EMZ plant.

Figure 3 shows an example of usable laminated cores and coils distribution as well as their electric connection (see figure 4). The main geometrical parameters are an outer diameter of 322 mm and an air gap width of 25.5 mm. The number of slots of the outer system is 48 and of the inner system 36.

Symmetrical time-dependent current distributions according to the phase connection are presented by the following equations (9):

$$i_A = \hat{I}\cos(\omega t), \qquad i_B = \hat{I}\cos(\omega t - \frac{2\pi}{3}), \qquad i_C = \hat{I}\cos(\omega t - \frac{4\pi}{3}), \quad (9)$$

where $i_{A,B,C}$ are the instantaneous values of electric currents in the phases A, B and C of the exciter system and $\hat{I}$ their peak value.

The calculation domain was split into uneven element sizes:

+ parts in the exciter systems with a large gradient of electromagnetic parameters, here the sizes of the cells are very small and
+ in the outer space with a small gradient of electromagnetic parameters, here the size of elements increases.

The mesh convergence presented in [6] as dependency of the error from the finite element number results in a low error of only 1.78 % at an element number of 13258. This result is suitable for simulation and all further simulations are carried out with this number of elements.

4. Results of the numerical simulations

The simulations of electromagnetic parameters are implemented for the following typical magnetic field strengths in the air gap of H_δ = 50, 70 and 100 kA/m with a frequency of the electric currents of f = 50 Hz.

To verify the obtained simulation results, the experimental measurement of magnetic flux density distribution in the middle of the air gap is carried out with the help of Hall Probe MNT-4E04-VH (Lake Shore) and Gauss meter 421 (Lake Shore). The results of these experimental measurements were compared with simulation data for the azimuthal and radial components. The relative differences amount up to maximal 7 %. This difference comes from the influence of the real magnetic properties and the geometry of the iron sheets, also from the discretization of the coil distribution.

The differences of the magnetic flux density magnitude B_{mag} obtained in ANSYS Maxwell and COMSOL Multiphysics are approximately 3 %. The differences between results in 2D and 3D simulation in all parameter distributions in the middle cross section reach 2-3 %.

The results of all comparisons attest that the developed numerical model is possible to be carried out in 2D and applicable for the electromagnetic design of exciter systems for electromechanical mills.

Figure 5 presents the radial and azimuthal component of the vector gradient of the magnetic field in the air gap of the manufactured EMZ plant on three circles, which are located in the near inner exciter system (R1), in the middle of the air gap (R8) and close to the outer exciter system (R16) at $H_\delta = 70$ kA/m and $f = 50$ Hz.

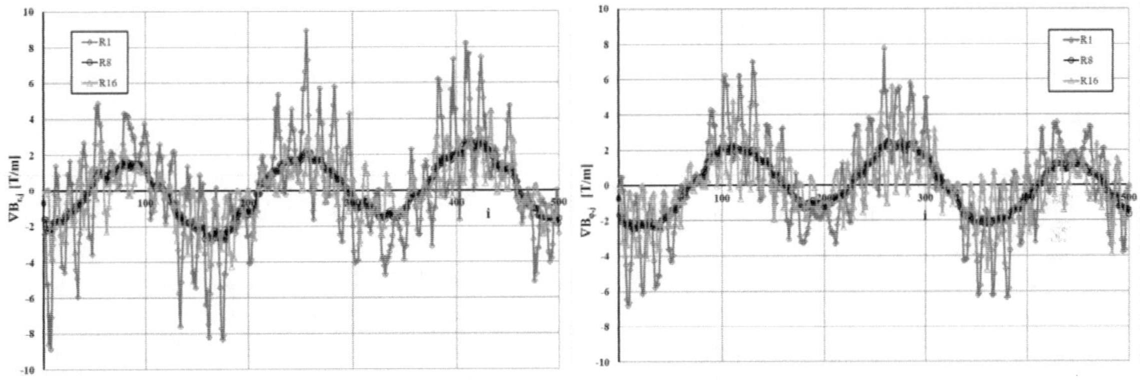

Figure 5. Radial $\nabla B_{r,j}(i)$ (left) and azimuthal $\nabla B_{\varphi,j}(i)$ (right) components of the vector gradient of the magnetic field in the air gap of the manufactured EMZ plant on three circles j, which are located in the near inner exciter system (R1), in the middle (R8) of the air gap, and close to the outer exciter system (R16) dependent on i at $H_\delta = 70$ kA/m, $f = 50$ Hz, $\omega t = 0$.

The components of the vector gradient of the magnetic field have the same order up to 8 T/m, but their changes are larger close to the exciter systems (circle R1 and R16) than in the center of the air gap (R8). Consequently, the milling beads are more accelerated or braked close to the exciter systems. Here the movement changes are maximal.

In the center of the air gap the distribution is approximately sinusoidal along the circle R8. However, they are shifted, so that the azimuthal component $\nabla B_{\varphi,j}$ is larger where the radial component $\nabla B_{r,j}$ is smaller or vice versa.

Other simulations [4] have shown that the magnitudes of the vector gradient components increase with the enhancement of the magnetic field strength H_δ in the air gap and the number of poles 2p as well as with the reduction of the number of slots per pole per phase (q $\rightarrow$ 1) and of the chording of the windings ($\varepsilon \rightarrow 0$). These are the potentials for further improvements of the performance of mills based on the electromechanical principle [7].

5. Dry and wet milling in developed laboratory plant

To check this approach, experiments in the laboratory installation (EMZ-LAI) with crushed cement clinker (dry milling) and anthraquinone (wet milling) were performed in dependence on the magnetic field strength H_δ at constant current frequency f and fillings rates of the working chamber.

The specific energy consumption w_0 of EMZ-LAI dependent on the milling ratio z is presented in figure 6 in comparison with a ball mill (Bond) and a stirred ball mill (RWKM) and the same educt (cement clinker < 500 µm).

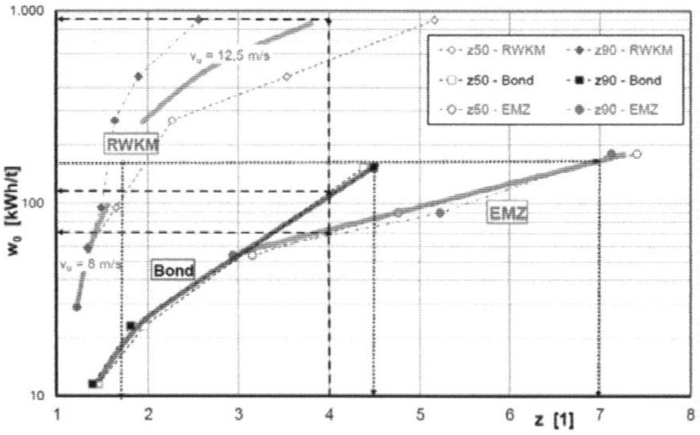

Figure 6. Specific energy consumption w_0 dependent on the milling ratio z of EMZ-LAI compared with a ball mill (Bond) and a stirred ball mill (RWKM with different stirrer speeds v_U) for dry milling of cement clinker < 500 µm.

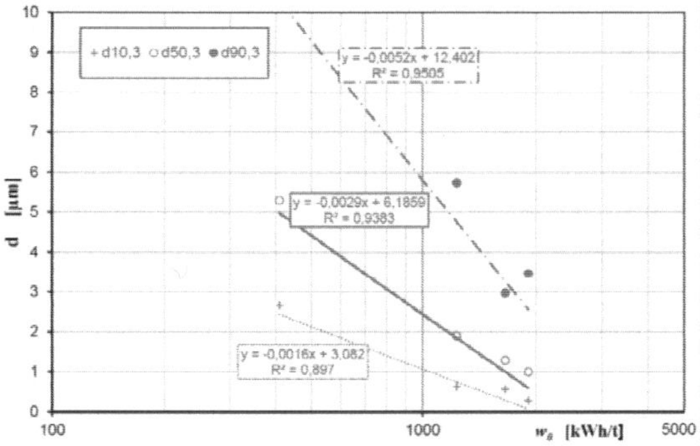

Figure 7. Particle sizes $d_{10,3}$, $d_{50,3}$, $d_{90,3}$ dependent on the specific energy consumption w_0 for wet milling of anthraquinone with input particle sizes of $d_{10,3}$= 17.1 µm, $d_{50,3}$ = 25.5 µm and $d_{90,3}$ = 37.1 µm in EMZ-LAI.

It is clearly derivable that the specific energy consumption of the EMZ-LAI is lower than w_0 of Bond and RWKM at the same milling ratio z (e.g. $z = 4$). The larger the required milling ratio of the product must be, the larger is the energetic advantage of EMZ principle. Furthermore, it is possible to achieve a high fineness (e.g. $z = 7$) only with the EMZ principle.

Figure 7 demonstrates the milling progress of a special aqueous model fluid which contains 5 wt-% anthraquinone particles (A90004-250G, Sigma-Aldrich Chemie GmbH / Taufkirchen) dependent on the specific energy consumption w_0. The results of these experiments show that the anthraquinone particles are electromechanically grinded to $d_{50,3}$ < 1 µm. However, the grinding progress for large particles ($d_{90,3}$; $d_{50,3}$) is faster than for small particles ($d_{10,3}$). The reason is the used size of the milling beads in the range of 1.0-1.6 mm. The implementation of larger grinding progress also for smaller particle sizes requires smaller milling beads < 1 mm.

6. Summary

Methods and approaches for numerical simulation of magnetic field and its vector gradient in ANSYS Maxwell and COMSOL Multiphysics are worked out. With the developed electromagnetic simulation tools on the basis of evaluation of the vector gradient distribution in the process room an up- and downscaling of EMZ plants is possible.

Furthermore, the possibility to confirm simulation results by experimental measurements and experimental milling results in the laboratory plant EMZ-LAI is shown.

The energy advantages for dry milling of building materials and wet milling of organic agents are shown using the example of cement clinker and using the example anthraquinone.

The next steps are the connection of the developed electromagnetic model with DEM-simulations [8] and with a stress model [9] to determine the required magnet field structure in the air gap for an efficient milling process.

References

[1] Merisko-Liversidge E M and Liversidge G G 2008 Drug Nanoparticles: Formulating Poorly Water-Soluble Compounds. *Toxicologic Pathology* **36** 43-48 DOI: 10.1177/0192623307310946

[2] Schäfer St and Hoenig V 2015 Chancen und Grenzen der energetischen Optimierung und CO_2-Reduktion in der Zementindustrie. *2. Energiepanel Energieeffizienz und CO_2-Reduktion in der Industrie* Mainz 25.03.2015

[3] Macko M Size 2012 Size Reduction by Grinding as an Important Stage in Recycling, *In: Post-Consumer Waste Recycling and Optimal Production* Edited by Enri Damanhuri Chapter **12** 273 - 294 DOI: 10.5772/2642

[4] Halbedel B and Kazak O 2017 Potenziale eines neuen Mahlverfahrens – die elektromechanische Zerkleinerung (EMZ) *In „Produktgestaltung in der Partikeltechnologie* Bd. 8 FRAUN HOFER Verlag Stuttgart 181-196

[5] Halbedel B, Ziolkowski M and Brauer H 2011 Potenziale des Elektromechanischen Zerkleinerungsprinzips. *In Proceedings: Workshop Elektroprozesstechnik*, Technische Universität Ilmenau, Ilmenau/Ortsteil Heyda 06.- 07. Oktober 2011 11

[6] Kazak O and Halbedel B 2015 Approaches for modelling electromagnetic parameters in milling demonstration plant of organic agents *In: Proceedings 8th International Conference Electromagnetic Processing of Materials* Cannes / France 12-16 October 2015 4

[7] Halbedel B and Kazak O 2017 Erregersystem für elektromechanische Mahlaggregate zum Zer kleinern, Desagglomerieren, Dispergieren und Mischen von dispersen Stoffen und pumpfähigen Mehrphasengemischen. DE 10 2017 008 513.7 23

[8] https://www.cfdem.com/liggghts-open-source-discrete-element-method-particle-simulation-code, 26.10.2017

[9] Kwade A and Stressing A 2003 A Stressing Model for the Description and Optimization of Grinding Processes. *Chemical Engineering & Technology* **26** (2) 199 – 205 DOI: 10.1002/ceat.200390029

Acknowledgment

The authors acknowledge financial support by the German Research Foundation (DFG) (Ha 2338/1) and by the Federal Ministry for Economic Affairs and Energy (BMWi) in ZIM-Cooperation Project EMZ-W (KF2184744KO4) as well as support by several industrial partners.

IOP Conf. Series: Materials Science and Engineering **355** (2018) 012022 doi:10.1088/1757-899X/355/1/012022

Numerical investigation of slag formation in an entrained-flow gasifier

G Zageris, V Geza, A Jakovics

Department of Physics, University of Latvia, LV-1002 Zellu iela 23, Riga, Latvia

girts.zageris@fizmati.lv

Abstract. A CFD mathematical model for an entrained-flow gasifier is constructed – the model of an actual gasifier is rendered in 3D and appropriately meshed. Then, the turbulent gas flow in the gasifier is modeled with the realizable k-ε approach, taking devolatilization, combustion and coal gasification in account. Various such simulations are conducted, obtaining results for different air inlet positions and by tracking particles of varying sizes undergoing devolatilization and gasification. The model identifies potential problematic zones where most particles collide with the gasifier walls, indicating risk regions where ash deposits could most likely form. In conclusion, effects on the formation of an ash layer of air inlet positioning and particle size allowed in the main gasifier tank are discussed, and viable solutions such as radial inlet positioning for decreasing the amount of undesirable deposits are proposed. We also conclude that the particular chemical reactions that take place inside the gasifier play a significant role in determining how slagging occurs inside a gasifier.

1. Introduction

Although biomass gasification is a well known process, producers of gasifier equipment still face certain problems. Ash melting and deposition phenomena are important problems which cause the formation of a slag layer on equipment walls and may lead to a reliability problem due to negative effects on wall heat transfer and chemical corrosion [1]. Furthermore, slagging is an important phenomenon, but insufficiently investigated both experimentally and numerically. A number of papers describe underlying processes, but linking between local variables of slag (e.g. slag thickness, slag growth speed, slag movement under gravity forces etc.) and global variables of the gas flow is still missing. Most modeling attempts are limited to one or two-dimensional models [2], thus missing spatial behavior of slagging process. This paper aims to explain 3-D effects of particle size and flow characteristics on the formation of slag on the walls of an entrained-flow gasifier.

2. Description of the model

2.1. Geometry

The mathematical model is built on an existing, approximately eight meters high industrial entrained flow gasifier, with dimensions, inlet pipe positions and inlet parameters taken from the technical specification of the gasifier. The gasification process is modeled in 3D. There are two types of geometries being analyzed – the whole gasifier in its entire height, and only the region where the inlet pipes enter the gasifier. This is done to analyze the inlet zone more in detail, as slagging occurs most intensely in that region.

Content from this work may be used under the terms of the Creative Commons Attribution 3.0 licence. Any further distribution of this work must maintain attribution to the author(s) and the title of the work, journal citation and DOI.
Published under licence by IOP Publishing Ltd

Additionally, to determine how inlet pipe positioning affects slagging tendencies, a few variations of the real inlet zone geometry are constructed, with inlet pipes at various angles, as well as a tilted variant. The different 3D models can be seen in Figure 1. For all cases, in units of bottom inlet diameter (i.e. the diameter of the tank at the bottom is 1), the total height is 2.09, the secondary inlet pipe diameters are 0.065, the inlet pipes connect to the main gasifier tank at a height of 0.83. The top section has a diameter of 1.94 units. The tank begins to widen at a height of 1.11.

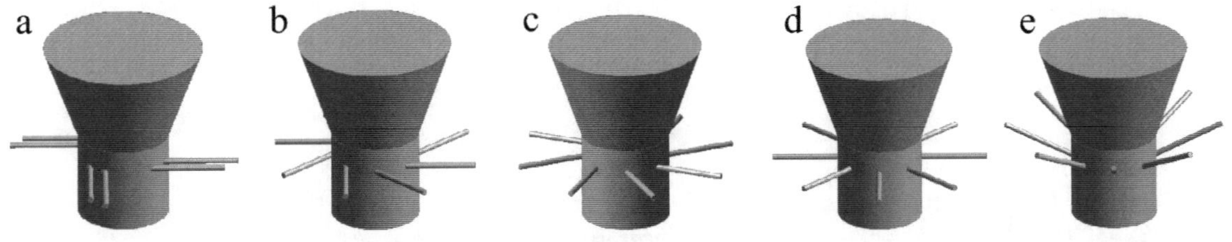

Figure 1. Gasifier geometries with different inlet pipe positioning. (a) - actual gasifier setup (b) - uniform inlets tangentially to the wall, (c) - uniform inlets turned 45 degrees toward the wall, (d) – uniform radial inlets, (e) uniform radial inlets facing downward by 30 degrees.

It should be noted that the geometry constructed is only valid for entrained-flow gasifiers – other types of gasifiers, such as the updraft or fluidized bed variants, are built differently, and therefore the results obtained in this paper would not pertain to gasifiers that differ strongly in principle from entrained-flow gasifiers. The results are valid only as long as the oxidizing agent and the biomass are fed co-currently through the gasifier, though, as we note later, the particular geometry and chemical reactions taking place are important to the result, even if the gasifier type matches with the one considered in this paper.

2.2. Gas mixture and flow

The gas inside the gasifier is modelled as an incompressible ideal gas. The flow regime of the gas mixture inside the gasifier is turbulent, and the realizable k-ε turbulence model is used to account for such behaviour in the model. In the *k-ε* model, two governing equations are solved – the turbulent energy equation (for k) and the turbulent dissipation equation (for ε) – in addition to the Navier-Stokes equation. Furthermore, heat transport is also modeled via the energy transport equation. The equations are solved numerically in ANSYS Fluent software in a mesh consisting of approximately 800k elements, with the number slightly varying over different geometries.

At the bottom of both models, a gas mixture (CO – 45%, N_2 – 26%, CO_2 – 16%, H_2O – 6%, CH_4 – 6%, H_2 – 1%) at 873 °C temperature is introduced at a constant mass flow rate consistent with operating conditions. The composition of this mixture was obtained from a separate calculation. There is a stage the biomass undergoes before entering the gasification chamber – first, it experiences a stage of devolatilization. This process was calculated separately, and the resultant gas mixture was then applied to the main model. This gas flow drives the co-current flow of the gasifier. Additionally, from every side inlet, a mixture of air and steam is introduced at 1073 °C, also at a constant mass flow rate. The mixing of these two flow sources and the resultant velocity field determines the behavior of particles as they travel through the gasifier.

As thermal effects are important for the process of gasification, the energy equation is also enabled. Thus, heat transfer due to diffusion, species transfer, chemical reactions etc. are taken in account. The heat flux through the gasifier walls is assumed to be zero.

2.3. Particles and gasification

Particles are introduced into the gasifier via the bottom at a constant mass flow rate. To investigate the role of particle sizing in slagging tendencies, four particle sizes were chosen – 5 μm, 100 μm, 500 μm and 1 mm in diameter. The particles are made up of 80% coal and non-combustibles. The process of

IOP Conf. Series: Materials Science and Engineering **355** (2018) 012022 doi:10.1088/1757-899X/355/1/012022

devolatilization takes place before the particles enter the gasifier, so volatile fraction is low. For all simulations, the mass flow rate of each type of particle is the same.

Because there are no more than a few hundred thousand particles inside the transition zone at any time, the particles constitute a very small volume fraction (no more than 0.05%) and thus the particle trajectories can be modelled using the Lagrangian approach.

The forces governing particle trajectory are as follows: inertial forces, gravity (facing downwards in the geometry sketches), thermophoretic force and a turbulence random walk force (with the characteristic time scale taken 0.30k/ε). The thermophoretic force manifests itself for particles smaller than a millimeter and is of the form:

$$\vec{F} = -D\frac{1}{mT}\nabla T \tag{1}$$

Here, D is the thermophoretic constant, m is the mass of the particle, T is the temperature around the particle, and ∇T is the gradient of the temperature.

The dominant forces in trajectory determination are the forces of inertia and gravity. The thermophoretic force adds small corrections, and the turbulence random walk model is enabled to account for small fluctuations of the turbulent gas flow that disappear when doing numerical calculations over averaged time steps – it adds a component of velocity to the mean velocity with a set of randomly generated normally distributed variables:

$$u_i' = \zeta_i\sqrt{\overline{u}_i^2} \tag{2}$$

Here, u_i is the i-th component of the fluctuation velocity, ζ is the random variable for that component of velocity, and the remaining term is the RMS of the i-th velocity component. The random variables are generated anew every time the fluctuations are applied (which is over a period equal to the time scale mentioned previously).

While the particles make their way through the gasifier, they also undergo chemical reactions that produce syngas. Additionally, there are reactions that take place in the volume of the gas as well. All reaction rates are modelled with the Arrhenius equation. The reactions taken in account are summarized in Table 1.

The particles are decoupled from the flow – first, the flow fields are calculated, then the particles trace through the acquired flow.

Table 1. Summary of chemical reactions implemented in the model.

Reaction type	Reaction equation	A	E, J/kmol	Reference
Volumetric	$CO + 0.5O_2 \rightarrow CO_2$	2.239×10^{12}	1.67×10^{8}	[3]
Volumetric	$CO + H_2O \rightarrow CO_2 + H_2$	9.87×10^{8}	3.1×10^{7}	[4]
Volumetric	$CO + 3H_2 \rightarrow CH_4 + H_2O$	5.12×10^{-14}	2.73×10^{4}	[3]
Volumetric	$H_2 + CO_2 \rightarrow CO + H_2O$	1.785×10^{12}	3.26×10^{8}	Equilibrium with reaction #2
Volumetric	$CH_4 + 1.5\ O_2 \rightarrow CO + 2H_2O$	5.012×10^{11}	2×10^{8}	[5]
Volumetric	$CH_4 + H_2O \rightarrow CO + 3H_2$	5.922×10^{8}	2.09×10^{8}	[6]
Surface	C<s> + $0.5O_2 \rightarrow CO$	300	1.3×10^{8}	[7]
Surface	C<s> + $CO_2 \rightarrow 2CO$	2224	2.2×10^{8}	[7]
Surface	C<s> + $H_2O \rightarrow CO + H_2$	42.5	1.42×10^{8}	[7]
Surface	C<s> + $2H_2 \rightarrow CH_4$	1.62	1.5×10^{8}	[7]

3. Results and discussion

3.1. Particle behaviour with no gasification reactions

To first see the isolated effect of the flow inside the gasifier tank on particle collisions with the walls of the gasifier, calculations that omit chemical reactions were made. These calculations were made on all geometry variants to see how inlet positioning can determine critical zones where slag deposition is most intense. The obtained flow velocity fields can be seen in Figure 2.

When the inlets are positioned tangentially with respect to the walls, the flow tends to swirl in the radial direction, but is mostly straight in the axial direction. In this regime, particles experience a centrifugal force as they travel upwards through the gasifier, which plays a role in particle collisions with the walls. When inlets are radial, secondary vortices appear in the axial cross-section. Here, the particles that sediment on the walls are those that are pulled inside these vortices and guided towards the walls.

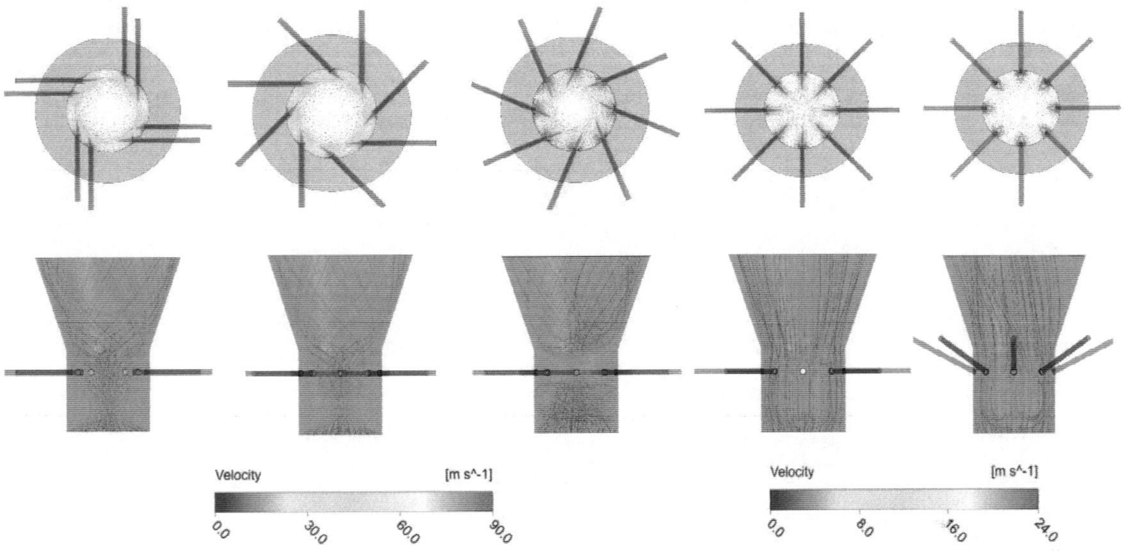

Figure 2. Flow velocity fields for all geometry variations – from above (top) and from the side (bottom). The left side scale is for the first three cases, the right-side scale is for the latter two cases.

Next, particle behaviour in all models was analysed. A simple wall collision model was used – every particle that hit a wall was terminated, and the position and size of the particle was reported. The obtained data is summarized in Table 2 and Figure 3.

Table 2. Results for particle collision in the model with no chemical reactions.

Model (refer to Figure 1)	**a**	**b**	**c**	**d**	**e**
Total collisions	1450000	1591463	1776187	1953891	758868
Total collisions, normalized	1.91	2.10	2.34	2.57	1
5 μm, %	5.3	9.5	10.7	2.4	2.8
100 μm, %	26.6	35.1	35.5	13.9	11.0
500 μm, %	47.5	39.6	36.9	47.6	38.3
1 mm, %	20.7	15.8	16.9	36.1	47.9

From the simulations, it is evident that the problematic slagging zones become more pronounced as the inlets are made more radial. In particular, a spike in collisions occurring immediately below the inlets arises once the inlets become radial (Figure 3 d and e). It is also notable that for tangential inlets (Fig 3. a and b) the collision zones are broad, owing to the centripetal forces that push transiting particles toward the walls. As the inlets are turned more radially, the broad impact zones tend to become sharper and center around the inlet zones due to secondary vortices that form immediately below the inlets and push particles toward the problematic zone.

Furthermore, larger particles (500 µm and 1 mm) generally tend to collide with the walls at low heights more than the smaller particles. This is especially evident in the tangential inlet configuration (Fig 3. a, b, c). As the heavy particles travel through the gasifier slower than smaller ones, they are exposed to the centrifugal forces for a longer time, allowing for collisions. Conversely, the 100 µm particles collide the least relative to the rest (see Table 2) – this is because the small particles can pass the inlet zone before colliding with the walls at all.

An interesting dynamic occurs when the radial inlets are tilted (Figure 3. e) – the total amount of particles experiencing collisions is the lowest of all possible configurations, however, the collision zone is very pronounced. In other words, there is a potential risk of forming a slag ring. To fully argument whether slagging occurs at the walls, though, gasification reactions must be taken in account and a more sophisticated trapping condition must be formulated for the particles.

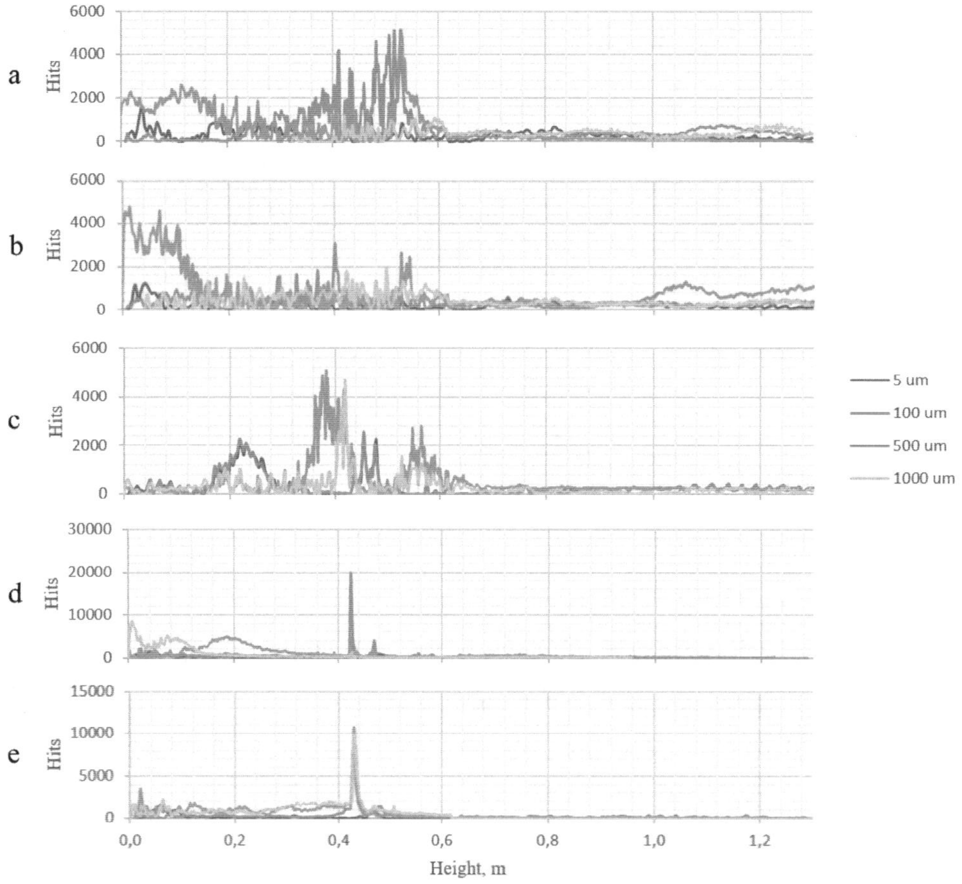

Figure 3. Histogram of amount of particles hitting the walls depending on the height of impact (numeration consistent with Figure 1).

3.2. Particle behaviour with no gasification reactions

Next, the full gasification model is enabled, allowing the gas and particles to chemically react with the gases present as described previously. Also, a more adequate trapping condition is formulated, based on the fusion temperature of the particles and conversion rate, which is calculated as follows [10]:

$$X = 1 - \frac{C_{char}^{carbon} C_{coal}^{ash}}{\left(1 - C_{char}^{carbon}\right) C_{coal}^{carbon}} \tag{3}$$

where X is the conversion rate of the particle and C is the weight fraction of the material in the top index in the bulk of the material in the lower index.

The condition depends on the state of both the particle and the wall [8,9,10]. The particle can be in either a sticky state (its temperature above fusion temperature, particle conversion above critical value) or a non-sticky state. Also, the fate of the particle depends on the Weber number (*We)*, which shows the ratio between inertial forces in the fluid and surface tension:

$$We = \frac{\rho v^2 l}{\sigma} \tag{4}$$

Here, ρ is the density of the fluid, v is the velocity of the particle, l is the characteristic length of the particle and σ is the surface tension.

There are different scenarios depending on whether the Weber number is above or below a critical value (taken to be 1). Similarly, the wall is deemed sticky if it is above the fusion temperature of the particle or if the impeding particle is sticky. The subsequent scenarios are summarized in Table 3.

Table 3. Particle slagging conditions [8].

	Sticky particle		**Non-sticky particle**	
	We<1	*We*>1	*We*<1	*We*>1
Sticky wall	Slagging	Slagging	Slagging	Reflect
Non-sticky wall	Slagging	Reflect	Reflect	Reflect

With these conditions in place, a simulation was run. The flow fields were practically unchanged relative to those depicted in Fig. 2, allowing to conclude that gasification processes do not dramatically change the flow characteristics. However, particle collisions are vastly different. The amount of collisions decreases rapidly in this model, and the tangential models report approximately 100 times more collisions than radial models as shown in Table 4. To explain this, temperature fields of cases a, d and e are shown in Fig. 4.

Table 4. Results for particle collision in the model with enabled chemical reactions.

Model (refer to Fig. 1)	**a**	**b**	**c**	**d**	**e**
Total collisions	987644	996731	769458	14842	10097
Total collisions, normalized	97.82	98.72	76.21	1.47	1
5 μm, %	5.3	9.5	10.7	2.4	2.8
100 μm, %	26.6	35.1	35.5	13.9	11.0
500 μm, %	47.5	39.6	36.9	47.6	38.3
1 mm, %	20.7	15.8	16.9	36.1	47.9

For the tangential cases, temperature spikes are located near the walls. In accord with the particle sticking condition, this makes particles prone to slagging. Also, as previously explained, large

IOP Conf. Series: Materials Science and Engineering **355** (2018) 012022 doi:10.1088/1757-899X/355/1/012022

particles are mostly forced towards the walls in the tangential configurations due to centripetal forces, further increasing the risk of excessive slag with large particle aggregation. The temperature spikes around the inlets and close to the walls are also reported experimentally, indicating that the model gives results that coincide with reality and giving a certain degree of validation.

Conversely, in the radial cases, the temperature spikes are in the middle, away from the walls. This gives an inverse effect – the gas temperature near the walls is relatively low, predominantly cooling the wall and particles, tending them towards non-sticky scenarios.

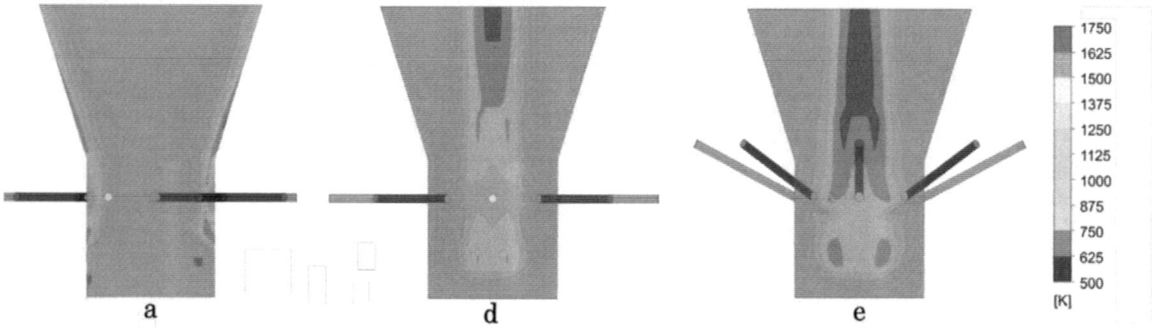

Figure 4. Temperature fields for various models with enabled gasification reactions.

It is important to take in account gasification reactions, as they determine the heat distribution inside the gasifier. As just shown, particle and wall temperature plays a large role in determining whether slag will form, overriding the tendencies appearing just from the analysis of the gas flow with no reactions. Thus, the probability of excessive slag deposition is also dependent on the particular geometry of the gasifier, though it appears that in axially symmetrical cases tangential inlets lead to temperature spikes near walls, significantly increasing the danger of slagging. To avoid slagging, a radial configuration for inlets is recommended.

4. Conclusions

A model for gasification in an entrained-flow gasifier was created and run in two modes – without the gasification chemical processes taking place, and then with the processes taken in account. The first model was used to analyze the impact of the flow inside the gasifier on the particles passing through it. It was determined that tangentially positioned inlets create a swirling flow that gives rise to centripetal forces that push particles towards the walls. Smaller particles experience this less as they travel through the gasifier quickly, but larger particles tend to collide with the walls because of this. In radial inlet configurations, a secondary vortex arises that pushes particles into a zone directly below the inlets.

In the second simulation run, gasification reactions are enabled and a more sophisticated particle capturing condition is formulated. With these in place, a dramatic change is reported – particles collided with walls far more in tangential configurations than radial ones. This is explained by the temperature fields in the gasifiers – in tangential cases, the largest temperatures are near the walls, while for radial cases the heat spikes are located in the middle. Therefore, a tentative recommendation to reduce excessive slagging can be made - slagging is decreased for gasifiers with axial symmetry if the inlets are positioned radially as opposed to a tangential positioning.

Finally, it is determined that chemical reactions inside the gasifier take a predominant role in determining slagging processes and potential risk regions, taking precedence over the flow profile inside the gasifier. Thus, it is important to take chemical processes in account when modelling slagging due to gasification.

Acknowledgements
This research was done with the financial support of European Regional Development Fund, project "Development, optimization and sustainability evaluation of smart solutions for nearly zero energy buildings in real climate conditions" (1.1.1.1/16/A/192).

References

[1] Lu X and Wang T 2015 Simulation of ash deposition behavior in an entrained flow coal gasifier, *International Journal of Clean Coal and Energy* **4** 43-59
[2] Bhuiyan A A and Naser J 2015 Modeling of slagging in industrial furnace: a comprehensive review *Procedia Engineering* **105** 512-19
[3] Westbrook C L and Dryer F L 1981 Simplified reaction mechanisms for the oxidation of hydrocarbon fuels in flames" *Combust. Sci. Technol.* **27** 31-43
[4] Bustanmante F, Enick R M, Killmeyer R P, Howard B H, Rothenberger K S, Cugini A V, Morreale B D and Ciocco M V 2005 Uncatalyzed and well-catalyzed forward water-gas shift reaction kinetics *AIChE J.* **51** 1440-54
[5] Gomez M A, Porteiro J, Patino D and Miguez J L 2014 CFD modelling of thermal conversion and packed bed compaction in biomass combustion *Fuel* **117** 716-32
[6] Hou K and Hughes R 2001 The kinetics of methane steam reforming over a Ni/α-Al_2O catalyst *Chem. Eng. J.* **82**, 311-28
[7] Wu Y, Smith P J, Zhang J, Thornock J N and Yue G 2010 Effects of turbulent mixing and controlling mechanisms in an entrained flow coal gasifier *Energy Fuels* **24 (2)** 1170-75
[8] Yong S Z and Ghoniem A 2012 Modeling the slag layer in solid fuel gasification and combustion – two-way coupling with CFD *Fuel* **97** 457-66
[9] Ni J, Yu G, Guo Q, Zhou Z and Wang F 2011 Submodel for predicting slag deposition formation in slagging gasification systems" *Energy Fuels* **25** 1004-09
[10] Li S, Wu Y and Whitty K J 2010 Ash deposition behavior during char-slag transition under simulated gasification conditions *Energy Fuels* **24** 1868-76

IOP Conf. Series: Materials Science and Engineering **355** (2018) 012023 doi:10.1088/1757-899X/355/1/012023

Benefits of magnesium wheels for consumer cars

Vilnis Frishfelds, Andrejs Timuhins, Uldis Bethers
Faculty of Physics and Mathematics, University of Latvia, Zellu 25, Riga LV-1002, Latvia
E-mail: frishfelds@latnet.lv

Abstract. Advantages and disadvantages of magnesium wheels are considered based on a mechanical model of a car. Magnesium wheels are usually applied to racing cars as they provide slightly better strength/weight ratio than aluminum alloys. Do they provide notable benefits also for the everyday user when the car speeds do not exceed allowed speed limit? Distinct properties of magnesium rims are discussed. Apart from lighter weight of magnesium alloys, they are also good in dissipating the energy of vibrations. The role of energy dissipation in the rim of a wheel is estimated by a quarter car model. Improvements to safety by using the magnesium wheels are considered. Braking distance and responsiveness of the car is studied both with and without using an Anti Blocking System (ABS). Influence of rim weight on various handling parameters of the car is quantitatively tested.

1. Introduction

Magnesium wheels have been frequently applied to racing cars as every reduced gram of the vehicle improves its lap times. This results from slightly better strength/weight ratio of magnesium alloys [1] in comparison with aluminum alloys. The weight of rim can be reduced by about 1/3 with respect to aluminum rim. Even better strength/weight ratio can be achieved by such metal processing technique as forging instead of less costly casting. Strong and fire proof Mg alloys are obtained with addition of Si, Mn, Al, Zn, Zr and other elements. There are also more costly Mg alloys with rare earth elements [2] included to get even better mechanical properties. But novel aluminum alloys are not much behind as some of them have nearly similar strength/weight ratio. Rim constitutes only a part of the total suspension weight where tire, axle and brakes contribute, too. Magnesium wheels of racing cars are used not only on tarmac where frequent acceleration, deceleration and cornering takes place but also for non-paved roads where car grip with uneven surface is essential [3]. One more distinct feature of magnesium is a strong dissipation of vibrations [4]. However, most of vibrations of the car are absorbed by the suspension and tires [5, 6]. Notable dissipation of vibration in the rim can occur only for very high frequencies at > 1 kHz that are not in characteristic range of oscillations of consumer cars. To evaluate these vibrations quantitatively we will consider rim deformation in a quarter car model [7]. Another feature of magnesium rims is efficient heat transfer that reduces heat pressure on brake disks, brake pads and brake drums in comparison with steel rims. However, heat conductivity of aluminum rims is approximately similar. Magnesium rims still has advantage as it heats up less than aluminum rim while braking but serious consequences are possible if it overheats. Magnesium rims provides unusual aesthetics of the car standing out from the crowd. That is key point for customers looking for alternatives of aluminum rims. It is more significant for consumer cars to have extra safety rather than to increase maximal speed

Content from this work may be used under the terms of the Creative Commons Attribution 3.0 licence. Any further distribution of this work must maintain attribution to the author(s) and the title of the work, journal citation and DOI.
Published under licence by IOP Publishing Ltd

and acceleration rate. Most important factors for safety is minimal braking distance and better handling capabilities of the car. It is clear intuitively that braking distance depends minimally on rim properties as wheels account for a small fraction of the car on every road with or without ABS. However, better handling of the car is essential in critical situations like in Moose test. There, a better responsiveness of the car is a must have. Thus, a lighter wheel may bring certain benefits which we would like to evaluate.

2. Mathematical model of the car

2.1. Quarter car model

Quarter car model [8] has been frequently applied to modeling of car dynamics for analyzing influence of tires, axle, disks, springs, road response, etc. Quarter car model will be used also in our case as it considers the main characteristics required for analyzing the role of rim properties during car movement. Vertical coordinate is defined as z, see figure 1. Displacement from unloaded state L_i of a component (deformation) is

$$\triangle z_i = z_{i+1} - z_i - L_i, \tag{1}$$

where index i denotes component: 0 is road, 1 - tire, 2 - rim, 3 - axle (unsprung weight without wheel), 4 - sprung weight, see table 1. The equation of momentum becomes

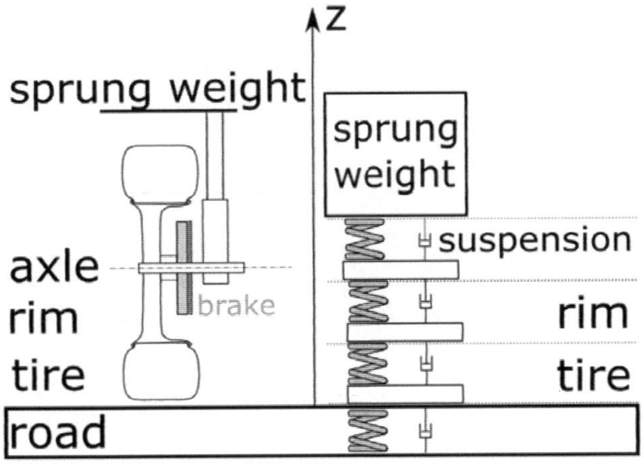

Figure 1. Scheme of the quarter car model.

$$m_i \frac{d^2 z_i}{dt^2} = m_i g + k_i \triangle z_i - k_{i-1} \triangle z_{i-1} - c_i \frac{d\triangle z_i}{dt} + c_{i-1} \frac{d\triangle z_{i-1}}{dt}, \tag{2}$$

where z_0 is road surface position. Variables k_i denote stiffness of the component and c_i - damping coefficients. Sprung weight is the last component of the system which is assumed to be rigid:

$$k_4 = c_4 = 0. \tag{3}$$

The roads are not always rigid but may deform under the weight of car, e.g. dirt roads. The quarter car model enables to add elasticity k_0 and damping factor c_0 for such roads. Parameters of the selected road are

$$k_0 = H(-\triangle z_0)\, k_{road}, \tag{4}$$

$$c_0 = H(-\triangle z_0)\, c_{road}, \tag{5}$$

Table 1. Properties of car components.

i	name	material	m, kg	L, m	k, N/m	ψ, -
1	tire		8.5	0.127	$2.25 \cdot 10^5$	0.2
2	rim	steel(Fe)	10	0.19	$1.28 \cdot 10^9$	0.005
2	rim	aluminum alloy(Al)	5.5	0.19	$4.42 \cdot 10^8$	0.02
2	rim	magnesium alloy(Mg)	4.5	0.19	$2.88 \cdot 10^8$	0.25
3	axle		25	0.35	$2.00 \cdot 10^4$	2.0
4	sprung mass		300	-	-	-

where $k_{road} = 10^7$ N/m, $c_{road} = 10^5$ N s/m and

$$H(x) = \begin{cases} 0 & x < 0 \\ 1 & x \geqslant 0 \end{cases} \tag{6}$$

is Heaviside step function. The Heaviside function takes into account that deformation of the road only occurs if wheel is in contact with the road. Characteristic parameters of a consumer car are given in the table 1. These data are derived from the best selling car model in the world, namely Toyota Corolla. Version of Corolla 2016 is used with base equipment and wheels. Damping coefficient of a component is derived from:

$$c_i = \frac{\psi_i}{2\pi}\sqrt{m_i k_i}, \tag{7}$$

where specific damping capacity ψ_i is ratio of energy lost and stored per cycle. Very high damping capacity is for suspension as it is accompanied by hydraulic or pneumatic shock absorbers improving ride quality. Damping of oscillations by the rim are usually disregarded but we will them include in order to estimate what this distinct feature of magnesium (table 1) can bring. Uneven road surface is assumed to be sinusoidal $z_0 = h_0 \sin(2\pi x/\lambda)$ with given wavelength λ and amplitude h_0. The period of car oscillations in road bumps is $T = \lambda/v$, where v is horizontal velocity of the car.

2.2. Friction model

Behavior of the car on the road depends on the grip with road surface. If the car has a good grip with the road it becomes more responsive to the drivers commands and shortens braking distance. That is the car is not sliding horizontally on the surface neither in the direction of the car motion nor in the normal direction. Horizontal coordinate in the direction of car motion is assumed to be x. Friction force is given by

$$F_x = \mu F_z, \tag{8}$$

where μ is coefficient of friction and F_z is the load of the tire to the road surface. The coefficient of friction $\mu = \mu(s)$ (see figure 2) can be estimated by Magic formula [9, 10] that depends on wheel slip ratio s [11]

$$s = \frac{v - \Omega R}{v}, \tag{9}$$

where Ω is the wheels rotational velocity, and wheel radius is $R = L_1 + L_2$.

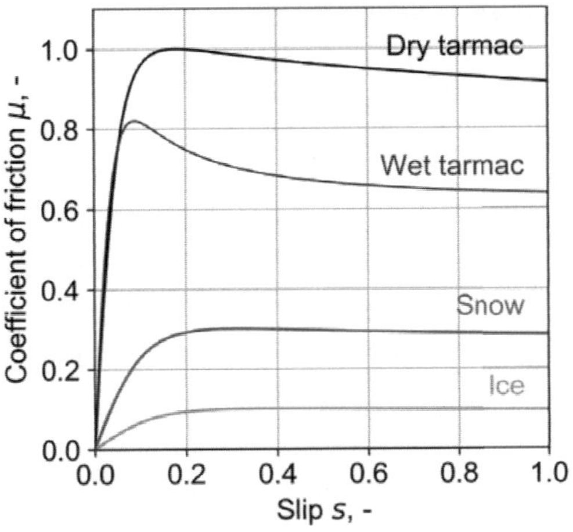

Figure 2. Tire longitudinal coefficient of friction for different road conditions [12].

Without wheels, the acceleration or deceleration rate of the car is given by

$$M\frac{d^2x}{dt^2} = -F_x, \tag{10}$$

where $M = \sum m_i$ is the quarter mass of the car. Wheel rotation must also be accounted, for, as wheels have non-zero inertial moment I:

$$I\frac{d\Omega}{dt} = rF_x - \tau, \tag{11}$$

where τ is braking torque, rF_x is road-tire friction torque with actual wheel radius $r = x_3 - x_1$. Moment of inertia I for the wheel with steel rim, aluminum rim and magnesium rim equal 1.215, 1.053 and 1.013 $kg \cdot m^2$, respectively. The difference is rather small because tire attributes have a dominating effect in moment of inertia. Tire load on the road is given by

$$F_z = c_0\frac{d\triangle z_0}{dt} - k_0\triangle z_0. \tag{12}$$

3. Frequency response of vibrating systems with quarter car model

Quarter car model is frequently applied to analyze car vibrations [8]. Oscillations are important for analyzing responsiveness of the car and driver comfort. Driver comfortably tolerates oscillations with periods of about one second. Therefore, the task of tires and springs is to reduce oscillations that have higher or lower frequencies as much as possible without notably compromising car responsiveness. Let us assume that harmonic oscillations occurs in the system with an angular frequency $\omega = \frac{2\pi}{T}$. Each component i of the car has an oscillations amplitude $|A_i|$ and phase $arg\,(A_i)$:

$$z_i = A_i \exp\,(j\omega t) \tag{13}$$

where $j =\sqrt{-1}$. Let us apply the oscillating external force to the tire as it is in real case where tire is forced by uneven road $z_1 = \exp\,(j\omega t)$ or $A_1 = 1 + 0j$.

The frequency response follows from the solution of linear system of equations (section 2):

$$-\omega^2 m_i A_i = k_i\,(A_{i+1} - A_i) - k_{i-1}\,(A_i - A_{i-1}) - j\omega c_i\,(A_{i+1} - A_i) + j\omega c_{i-1}\,(A_i - A_{i-1}) \tag{14}$$

for each component $i = 2, 3, 4$, see table 1. It is clear that a lighter wheel will follow the external oscillations more easily, i.e, it will read the road better. That we can see for the axle component in frequency diagram figure 4, where amplitudes of the quarter car components are shown excluding the tire which oscillates with a defined amplitude. Relative displacement of the rim almost coincides with the axle component for frequencies below 500 Hz, see figure 3 . For higher frequencies, there are resonance peaks for rim oscillations. Due to the strong damping of magnesium the resonance peak of magnesium is much weaker than for steel and aluminum.

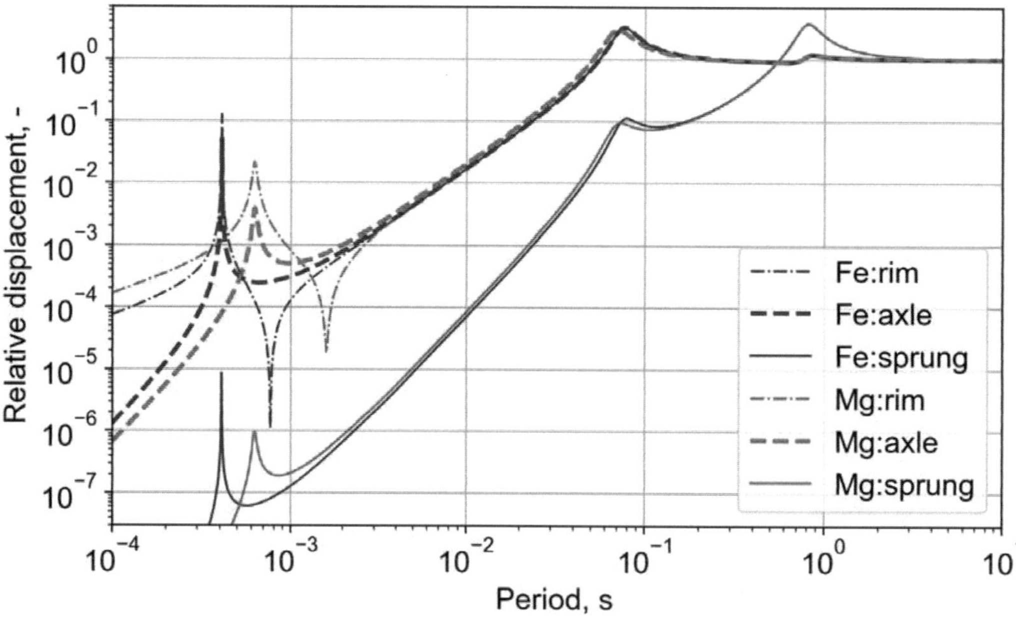

Figure 3. Relative amplitude of oscillating components $|A_i| / |A_1|$ on uneven road.

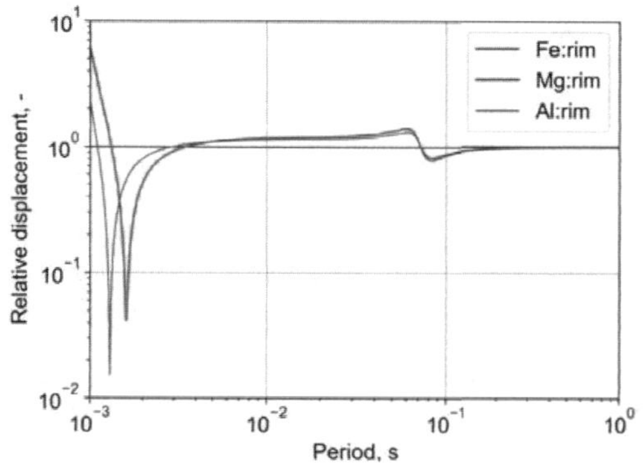

Figure 4. Relative amplitude of rim oscillations $|A_2| / |A_2^{Fe}|$ on uneven road.

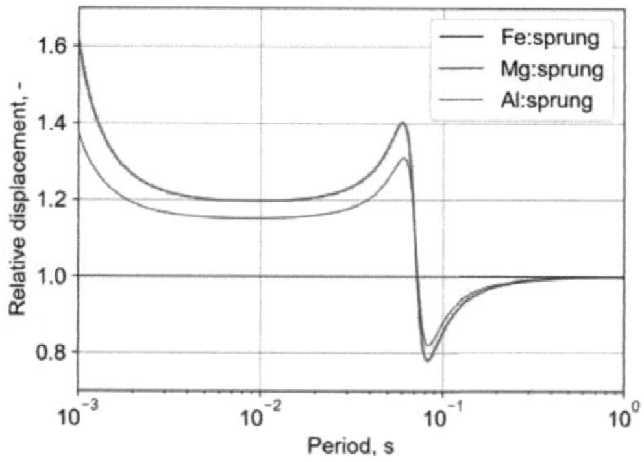

Figure 5. Amplitude ratio of sprung mass with respect to the case with steel rim $|A_4| / |A_4^{Fe}|$.

Oscillations of the car body, i.e. sprung weight are much smaller than wheel oscillations. As we can see in figure 5, there are also counter effect of lighter wheels: slightly increased car body vibrations, when driving on uneven surface with small wavelength $\lambda = vT$ of the bumps ($\lambda \approx 0.5$ m at speed $v > 50$ km/h), e.g. on gravel road, whereas it is opposite for longer vibrations ($\lambda > 1$ m at $v = 50$ km/h). It happens because tire of heavier wheel absorbs high frequency oscillations better. Nevertheless, the situation is opposite at longer bumps: car with steel rim will feel the bumps more as suspension with heavier wheel is closer to the resonance frequency of the spring. Lighter wheels help to increase the lifetime of tires as there are smaller tire oscillations for the same forcing by the road. Dampening factor of vibrations of magnesium rim is 10 times higher than of aluminum rim. However, figure 3 shows that notable damping by magnesium rim occurs at periods of less than a millisecond that are largely damped by tires on road with fine waviness. Therefore, damping of oscillations by the rim can be disregarded as amplitudes of axle and rim components almost coincide, see figure 4. Nevertheless, strong dampening increases durability of suspension brackets and brakes. The elastic and damping properties of springs have to be slightly adapted for given weight of the wheel in order to improve driver's comfort. This is, however, applied rarely for consumer cars.

4. Braking with and without ABS

One of the most important aspects of car safety is braking distance. The braking force can be well estimated with Magic formula, see sub-section 2.2 and (9). Do the lighter wheels improve braking distance? Let us test this on wet tarmac without ABS, i.e., when wheels can become blocked. Consider that initial velocity v_0 is 90 km/h, and the road is uneven with wavelength λ = 1 m and amplitude 1 cm. Such amplitude is insufficient to cause wheel of loosing contact with the road surface. The set of equations of section 2 are solved numerically for initial conditions $x|_{t=0} = 0$, $\frac{dx}{dt}|_{t=0} = v_0$ and $\Omega|_{t=0} = \frac{v_0}{2\pi r}$ in the time interval $[0, t_b]$, $Mv_0 = \int_0^{t_b} F_x dt$. Value of braking torque 1400 N·m is applied characteristic for hard braking. Figure 6 shows that braking distance is almost independent of wheel weight. The car with steel rim decelerates more at one side of road bump and the opposite at the other side. The average load on the surface remains weakly dependent on wheel weight. Thus, the braking distance is about the same for all rim materials .

Now, consider the braking by ABS. ABS enables to reduce braking distance on solid road without loosing the control of the car while braking. ABS is usually required in critical situations, when a sudden object appears on the road, that must be bypassed as it is in the moose test. Basically, ABS system detects wheel lockup, than commands to release brake pressure until the wheel rotation spins up [13]. Wheel lockup is usually defined as critical value of slip ratio s (9). Most effective and universal s is usually around 0.2 for all kinds of roads (see figure 2) when friction between road and tire is at maximum. Let the critical slip ratio be 0.3. The braking distance and time is clearly shorter then braking using ABS, see figure 6. Wheel rotation rate accelerates and decelerates multiple times during critical braking on wet tarmac. Inertial moment for light metal wheel is lower. Therefore, it has lower period between the cycles. On the other hand, lower period reduces time interval when the grip with road is reduced, i.e. when wheel has lowest rotational speed. The braking distance is few centimeters longer with magnesium rim and this difference is acquired in the final part of deceleration where road roughness comes into play, see figure 6. This difference can arise just from the phase of road roughness. Larger difference can only occur if the time interval between successive wheel lockups is comparable with duration required for braking. The speed of rotation with light metal wheel oscillates faster, that minimizes the time interval with reduced grip, see figure 7. Highest oscillations of deceleration rate are when the frequency of road at given speed is comparable with the resonance frequency of the tires. For the given road with wavelength $\lambda = 1$ m, the resonance occurs at velocities around 50 km/h. For higher velocities, road grip with heavier wheel oscillates even less because the road roughness is then absorbed more effectively. But the situation is opposite at lower car velocities with more steady deceleration rate of lighter wheel. It is clear that a

IOP Conf. Series: Materials Science and Engineering **355** (2018) 012023 doi:10.1088/1757-899X/355/1/012023

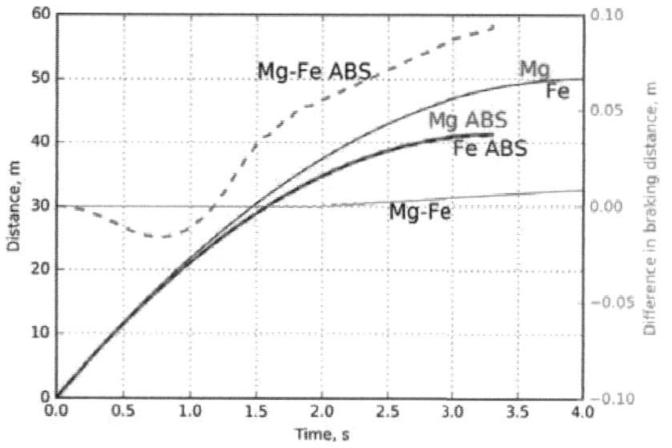

Figure 6. Braking distance on wet uneven tarmac with initial velocity 90 km/h for car with magnesium and steel rims both with and without ABS. Secondary axis shows difference between them (Mg-Fe).

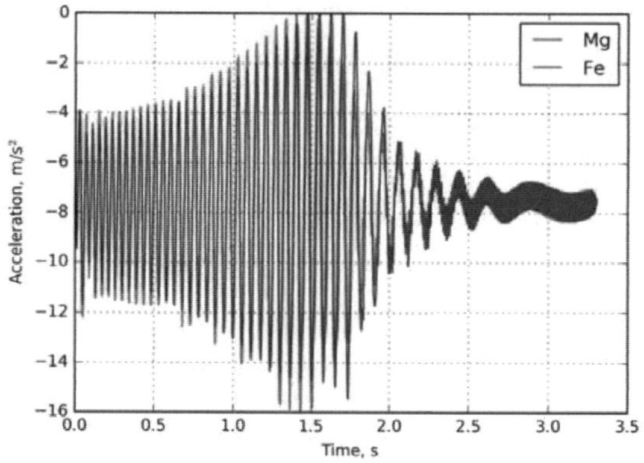

Figure 7. Deceleration rate of the car with ABS on wet uneven tarmac with initial velocity 90 km/h for car with magnesium and steel rims.

car with different rim weight requires different ABS settings [11]. Heavier wheel needs lower threshold s (9) as it is more difficult to accelerate it while braking and regain grip.

5. Weight advantages of magnesium wheels

Magnesium wheels have a lower weight than steel or aluminum wheels. This property leads to multiple advantages [8]:

(i) Better grip on uneven surface, as lighter wheels read the road better. As a result:
- It is easier to handle the car. Car motions are more predictable.
- Reduced risk of car sliding.
- Magnesium wheels allows higher speed for the same road without compromising grip with the road.

(ii) Increased durability of tires as amplitude of oscillations reduces in tire (see section above).

(iii) Easier handling of the car during emergency braking.

(iv) Lighter wheels improves acceleration and deceleration of the car on road with good grip, as the mass of the car and moment of inertia of the wheels is reduced. Considering that most of the rim weight is distributed at its exterior, it makes up to 2.2 % improvement for the car, see table 1, with magnesium rims in comparison with steel ones.

(v) Reduced mass of the car.

One drawback of lighter wheels is that the center of gravity of the car becomes slightly higher which affects car stability. Let us consider these effects quantitatively, assuming that an uneven road surface is sinusoidal with a given wavelength λ and amplitude.

5.1. Tire grip on road

Load of the tire on the road remains constant when driving on even surface and constant speed. Driving on an uneven road, the tire load on the surface is alternating but the average is the same as for even road. However, by increasing the vehicle speed, it is possible that the tire lose contact with the road surface and the car becomes irresponsible. Lighter wheels allow driving the vehicle with the same speed on more uneven road without compromising the grip of the car. Let us apply the quarter car model of section 2 to analyze responsiveness of the car. Quantitatively, good car grip means high and continuous load on the surface which leads to better response to drivers input, see figure 8 where minimal tire load on the surface is shown over a wavy road with amplitude of 1 cm. By increasing the speed of the car or decreasing the wavelength of the road surface it is possible to make the wheel lose contact with the road surface. Then it is not possible for the driver to control the car. For various wheels critical roughness for the road when this happens is shown in figure 9. The difference between magnesium and aluminum rims is relatively small but the car with steel rims needs almost twice smoother road. Two minimums of the critical roughness are associated with resonance frequencies of tires and suspension. If we move to higher frequencies above 12 Hz, then tire of the heavier wheel can absorb the road oscillations better. But small period of oscillations makes these vibrations less essential for handling of the car, as the driver cannot lose road contact in such a small interval of time. The car is easily handleable if the grip is almost continuous. Therefore, it is better

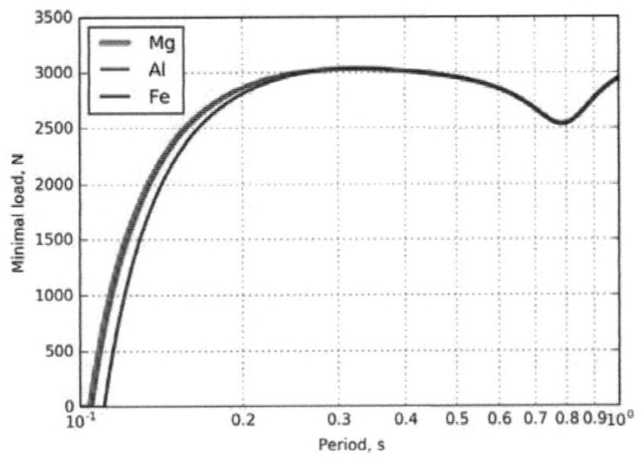

Figure 8. Minimal tire load vs. period $T = \lambda/v$ on uneven surface if the roughness amplitude is $h_0 = 1$ cm.

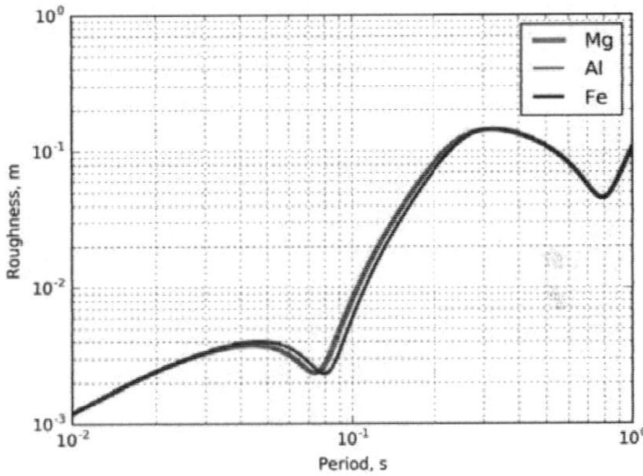

Figure 9. Critical roughness of the road vs. period $T = \lambda/v$ when wheel looses contact with the road.

that the amplitude of load oscillations is minimal. For example, driver is steering the wheels and the car responds weakly when the car load on the road is minimal. However, when the load oscillates to a higher value the driver's input can appear too large. Therefore, let us plot ratio of minimal vs. maximal load of tire on the surface, see figure 10. Again it is clear that magnesium and aluminum rims provide essential advantage over more heavy steel rims. More important is the duration of low grip of the car. Lighter wheel leads to shorter time interval with reduced grip at periods above 0.1 s, see figure 11. Reduced grip on the road can last about 0.01 s more for wheels with steel rims as compared to light metal rims.

6. Conclusions

Modeling results show that better dampening factor of magnesium wheels has minor influence in car handling. However, the lighter wheels substantially improves car responsiveness in many aspects. Also the tire lifetime is increased due to smaller oscillations in them. Braking distance is almost the same for any rim either with or without ABS but car responsiveness during critical braking is substantially improved by using combination of ABS and lighter wheels. One minor drawback of light metal wheels is higher car vibrations on road with fine waviness when slightly larger part of oscillations are transmitted to the sprung mass. Accounting the price of magnesium wheels and their durability, there are no significant advantages over aluminum wheels for every day user at velocities below 100 km/h. Most significant factor is that critical velocity is increased by a few percent on bumpy road when car still responds to driver. However, better look and

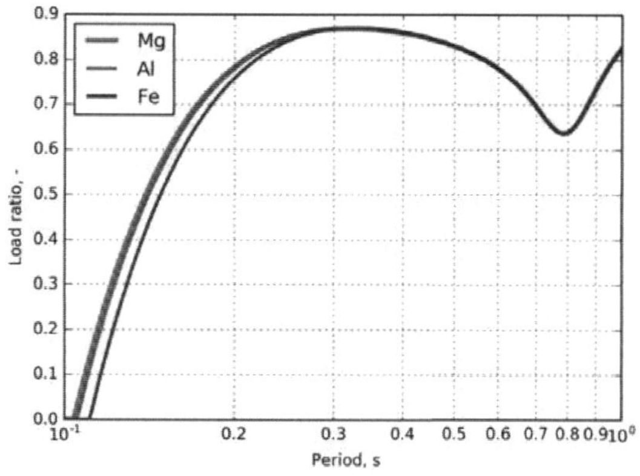

Figure 10. Ratio between minimal and maximal load on uneven surface vs. period $T = \lambda/v$ if the roughness amplitude is $h_0 = 1$ cm.

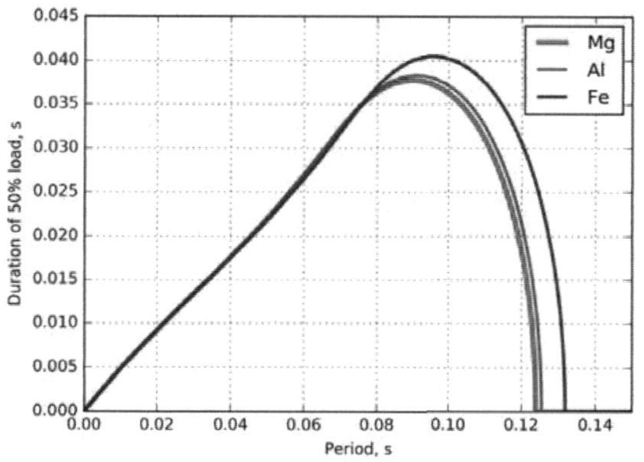

Figure 11. Time interval when the tire pressure on road is less than 50 % from normal vs. period $T = \lambda/v$ on road with roughness amplitude $h_0 = 1$ cm.

minimally better driving responsiveness will attract car enthusiasts especially for premium cars. Moreover, car acceleration rate increases by about 2 %. Most important benefits are for cars driving on an uneven surface especially with low profile tires. However, these wheels usually require larger rims where magnesium ones may appear to costly over aluminum ones if the lifetime of rims is included.

References

[1] Pitman-Dunn laboratory 1975 *High strentgh - high damping capacity wrought magnesium alloys* (Philadelphia: US Army Armament Command)
[2] Wang J, Wu Z, Gao S, Lu R, Qin D, Yang W, Pan F 2015 *Journal of Magnesium and Alloys* **3** 79
[3] Pacejka HB, Besselink I 2012 *Tyre and Vehicle Dynamics* (Amsterdam: Elsevier)
[4] Maringer RE 1966 *Damping capacity of materials, Volume I* (Alabama: US Army Missile Command)
[5] Clark SK ed. 2006 *The pneumatic tire* (Washington: US Department of Transportation)
[6] Koutny F 2007 *Geometry and mechanics of pneumatic tires* (Zlin: CZE)
[7] Wong YI 2001 *Theory of ground vehicles* (New York: John Willey & Sons, Inc.)
[8] Jazar RN 2008 *Vehicle Dynamics: Theory and Applications* (New York: Springer)
[9] Beckman M, Andreasson J 2003 *Proc. symposium of the International Association for Vehicle System Dynamics (Stockholm)* ed. Fritzson P pp 385-392
[10] Besselink IJM, Schmeitz AJC, Pacejka HB 2010 *Proc. International Modelica Conference (Linköping)* ed. Berg M, Trigell AS pp 337-352
[11] Pedro JO, Dahunsi OA, Nyandoro OT 2012 *Journal of Mechanical Science and Technology* **26** 4115
[12] Cerdeira-Corujo M, Costas A, Delgado E, Barreiro A 2016 *Proc. 12th Portuguese Conference on Automatic Control (Guimares)* ed. Garrido P, Soares F, Moreira AP pp 751-762
[13] Day TD, Roberts SG 2002 *A Simulation Model for Vehicle Braking Systems Fitted with ABS* (Warrendale: SAE technical paper series)